FIREFIGHTER WRITTEN EXAM STUDY GUIDE

NORTH CAROLINA
STATE BOARD OF COMMUNITY COLLEGES
LIBRARIES
SAMPSON COMMUNITY COLLEGE

BY

ARTHUR R. COUVILLON

INFORMATION GUIDES
HERMOSA BEACH, CALIFORNIA

FIREFIGHTER WRITTEN EXAM STUDY GUIDE

1990

BY ARTHUR R. COUVILLON

FIRST EDITION
ALL RIGHTS RESERVED.
THIS BOOK, OR PARTS THEREOF,
MAY NOT BE REPRODUCED IN ANY FORM
WITHOUT PERMISSION OF THE PUBLISHER
Printed in the United States of America
COPYRIGHT, 1990 BY INFORMATION GUIDES

Library of Congress Cataloging Data:

Couvillon, Arthur R.
"Firefighter Written Exam Study Guide"
1. Fire Service---Handbooks, manuals, etc.
2. Fire Engineer--Handbooks, manuals, etc.
3. Fire Captain---Handbooks, manuals, etc.
 4. Firefighter----Handbooks, manuals, etc
5. Information----Handbooks, manuals, etc.
6. Promotional----Handbooks, manuals, etc.
I. Title
LCCN 89-81736
ISBN 0-938329-59-6

INTRODUCTION

The intent of this "STUDY GUIDE" is to provide a convenient source of information pertaining to the knowledge and duties necessary for the position of: **FIREFIGHTER**.

The goal of this "STUDY GUIDE" is to provide the knowledge to any motivated individual desiring to secure the position of: **FIREFIGHTER**.

A vast number of Cities across the U.S.A. give exams for the position of: **FIREFIGHTER**. The information included in this "STUDY GUIDE" is presented in a unique way, so as to cause studying to be more organized and easier. As you read through this "STUDY GUIDE" you will come across some duplication of information, this information will be presented in a slightly different manner, as a study technique, which may help some individuals to retain certain bits of information.

Excluding information of a local nature and particular department policy, you will find that the study material covered in this "STUDY GUIDE" will be an excellent resource for you to use in studying for **ENTRANCE LEVEL FIRE DEPARTMENT WRITTEN EXAMS**. Included are enormous amounts of required knowledge and information that each individual will need in order to pass and increase the final score on the exam.

Remember most all Fire Department exams will have questions that will cover their own State and local laws, codes and ordinances along with the departments policy and procedures, rules and regulations, etc. Make sure that you cover these particulars for the department you are testing for!

ACKNOWLEDGEMENTS

Appreciation is expressed to those individuals, aspiring Firefighters, rookie and experienced Firefighters, Fire Engineers, Fire Captains, Training Officers, Fire Chiefs, and Fire Science Instructors who have contributed to the development of this "STUDY GUIDE".

A special thanks to Battalion Chief **PAUL STEIN**, of the Santa Monica Fire Department, for his part in the editing of this book!

ABOUT THE AUTHOR

The author is a veteran Firefighter, with 20 years of Firefighting experience. Art shares the knowledge that he has gained during this period with a series of published Fire Service Study guides:

A SYSTEM FOR BECOMING:
"THE COMPLETE FIREFIGHTER CANDIDATE"

"FIREFIGHTER WRITTEN EXAM STUDY GUIDE"

"FIREFIGHTER ORAL EXAM STUDY GUIDE"

A SYSTEM FOR: "ADVANCEMENT IN THE FIRE SERVICE"

"FIRE ENGINEER WRITTEN EXAM STUDY GUIDE"

"FIRE ENGINEER ORAL EXAM STUDY GUIDE"

"FIRE CAPTAIN WRITTEN EXAM STUDY GUIDE"

"FIRE CAPTAIN ORAL EXAM STUDY GUIDE"

All books are available from: INFORMATION GUIDES, P. O. Box 531, Hermosa Beach, CA 90254

TABLE OF CONTENTS

	PAGES
SECTION 1 : WHAT TO PREPARE FOR	**1-6**
SUBJECT MATTER	2
PRE TEST MANUALS	3
TYPES OF EXAMS	3-5
EXAM SUGGESTIONS	6
SECTION 2 : GENERAL APTITUDE AND JUDGEMENT	**7-19**
SITUATION QUESTIONS	8-19
SECTION 3 : READING COMPREHENSION AND VOCABULARY	**21-77**
EXPLANATION	22
READING QUESTIONS	22-64
VOCABULARY QUESTIONS	65-79
SECTION 4 : SPELLING AND GRAMMAR	**81-101**
SPELLING QUESTIONS	82-92
GRAMMAR QUESTIONS	93-101
SECTION 5 : MECHANICAL COMPREHENSION	**103-155**
EXPLANATION	104
GENERAL INFORMATION	105-107
TOOLS AND EQUIPMENT	108-109
MECHANICAL INFORMATION	110-118
MECHANICAL PRINCIPALS	119-120
QUESTIONS: GEARS/PULLEYS/LEVERS	121-144
MECHANICAL QUESTIONS	145-155
SECTION 6 : SCIENCE KNOWLEDGE CHEMISTRY AND PHYSICS	**157-179**
EXPLANATION	158
FIRE SCIENCE INFORMATION	159-164
FIRE HYDRAULICS	165-167
PHYSICS	167-171
QUESTIONS: SCIENCE/CHEMISTRY/PHYSICS	172-179

TABLE OF CONTENTS
(continued)

	PAGES
SECTION 7 : MATH CONCEPTS/ARITHMETIC	**181-200**
EXPLANATION	182
ADDITION PROBLEMS	183-184
SUBTRACTION PROBLEMS	185-186
MULTIPLICATION PROBLEMS	187-188
DIVISION PROBLEMS	189-190
PERCENTAGE PROBLEMS	191-192
WORD PROBLEMS	193-200
SECTION 8: FIRST AID/EMT TECHNIQUES	**201-215**
EXPLANATION	202
FIRST AID INFORMATION	202-203
FIRST AID QUESTIONS	203-207
EMT-I INFORMATION	208-210
EMT-I QUESTIONS	211-215
SECTION 9: PROGRESSIONS	**217-221**
EXPLANATION	218
ALPHABETICAL PROGRESSIONS	219
NUMERICAL PROGRESSIONS	220-221
SECTION 10: MATCHING FORMS AND PATTERN ANALYSIS	**223-243**
EXPLANATION	224
CUBES	225-229
PATTERNS/FIGURES	230-243
SECTION 11: GENERAL KNOWLEDGE	**245-312**
EXPLANATION	246
DRIVING	247-249
LADDERS	250-252
FIRE PUMPS	253-260
FIRE STREAMS/HOSE	261-267
FORCIBLE ENTRY/SALVAGE/VENTILATION	268-269
FIRE PREVENTION/BUILDING CONSTRUCTION	270-275
HAZARDOUS MATERIALS	276-284
FIREFIGHTING/FIRE BEHAVIOR	285-292
EXTINGUISHING SYSTEMS	293-301
SECTION 12: AFTER THE EXAM	**313-314**
FOR FUTURE REFERENCE	314
INDEX :	**317**

vi

SECTION 1
WHAT TO PREPARE FOR

SUBJECT MATTER

The written portion of the ENTRANCE LEVEL FIREFIGHTERS EXAM may cover several of the various categories of subject matter.

Each jurisdiction will have their own criteria for selection of subject matter. The subject matter will vary from one jurisdiction to another.

The categories of subject matter could include selections from the following list:
1. Aptitude.
2. Reading comprehension.
3. Verbal ability.
4. Vocabulary
5. Spelling.
6. Grammar.
7. Mechanical comprehension.
8. Mechanical ability.
9. Science.
10. Arithmetic.
11. Math and math concepts.
12. General information.
13. General knowledge.
14. First aid.
15. Emergency Medical Technician techniques.
16. Alphabetical progressions.
17. Number series/progressions.
18. Cubes, figures, matching forms.
19. Pattern analysis.
20. Intelligences.
21. Chemistry.
22. Physics.
23. Judgement.
24. Information included in an entry level Firefighter Candidate Preparation Manual: EXAMPLES
    a. Fire Chemistry.
    b. Fire hose.
    c. Fire Department ladders.
    d. Fire Department tools and equipment.
    e. Ventilation techniques.
    f. Overhaul techniques.
    g. Salvage techniques.
    h. Rescue practices.
    i. Fire Prevention.
    j. ETC.

Information regarding the above list will be covered in the following sections of this book, each of the sections in this "STUDY GUIDE" will cover, in detail, information pertaining to a particular subject contained in the above list.

ENTRY FIREFIGHTER PRE-TEST PREPARATION MANUALS

There are many jurisdictions that now use these manuals. They will usually hand them out when to you when they accept your application for employment.

These manuals may include any of the previously mentioned subject matter and/or they may contain various information that is directly related to the Fire Service, such as:
1. Driving/Apparatus and Apparatus Mechanics.
2. Tools and Equipment.
3. Fire Hose and Hose Streams.
4. Fire Pumps.
5. Fire Hydraulics.
6. Water Supply.
5. Ladders.
6. Ventilation, Overhaul, and Salvage.
7. Extinguishing Systems.
7. Fire Prevention.
8. Rescue Techniques and First Aid.
9. Fire Behavior/Chemistry.
10. Hazardous Materials.
11. Miscellaneous Information.

The above list will be covered in the following sections of this "STUDY GUIDE". Most of the information relating to the Fire Service will be covered in sections: 6 and 11.

These manuals are used in exams that will include a reading comprehension section that will assess your capability to retain information, recall the information, and then to appropriately interpret this information that has been presented in the manual.

There are some jurisdictions that will schedual information sessions on the material covered in the manuals that they hand out.

The serious candidate will attend various Fire Service related classes at local colleges so that he/she will have already attained much of the knowledge neccessary for exams of this type.

TEST INFORMATION WILL BE TAKEN DIRECTLY FROM THE INFORMATION PRESENTED IN THE MANUAL!

TYPES OF EXAMS

The subject matter included in the test is usually presented in the form of a "MULTIPLE-CHOICE EXAM".

MULTIPLE CHOICE EXAMS:

Require that you choose an appropriate answer from a list of answers that have been offered for a question that has been presented. (usually 4 or 5 choices)

EXAMPLE #1

The questions in a Firefighters multiple choice exam will be:

A. True-false.
B. Essay.
C. Multiple choice.
D. None of the above.

From the information that you have been given above, you know that "C" is the correct answer from the above choices.

EXAMPLE #2

The number of days in a year is:

A. 365.
B. 366.
C. 367.
D. 368.

The answer you should choose is choice A" because it is the one which is MOST OFTEN correct. Choice "B" is true for leap years, but most years have 365 days. Choice "A" is the BEST answer.

Some of the other types of entry level Firefighter WRITTEN EXAMS are:

1. True-false.
2. Fill in the blanks/completion questions.
3. Essay/short answer test.
4. Matching questions/answers.

TRUE-FALSE QUESTIONS:

Are questions that make a statement that is either true or false. You should respond by answering the statement as TRUE (correct) or FALSE (incorrect).

EXAMPLE:

QUESTION: A Firefighter that is deficient in judgement is a Firefighter that is lacking in the quality of making wise decisions. TRUE or FALSE ?

ANSWER: TRUE.

FILL IN THE BLANK TYPE QUESTIONS:

Are the type that will ask you to fill in a missing word or words so as to correctly complete a sentence or a statement. These type of questions are more difficult to answer, since guessing is eliminated. No answer choice is given, you must recall from your memory the correct answer.

EXAMPLE:

QUESTION: A fire _____ is a model of the four elements required by a fire: fuel, heat, oxygen, and uninhibited chain reaction, each side is contiguous with the other three.

ANSWER: Tetrahedron.

ESSAY QUESTIONS:

Are the type of questions that will require a written account to a question or statement. NOTE: these type of questions are rarely used in entry level exams, because they require a great deal of time to score.

EXAMPLE:

QUESTION: In the Fire Service what does the term FLASHOVER refer to?

ANSWER: Flashover refers to the stage of a fire at which all surfaces and objects are heated to their ignition temperature and flame breaks out almost at once over the entire surface.

MATCHING QUESTIONS:

Are the type of questions that will contain columns of material/information (usually two columns) which contain facts, dates, statements, numbers, ETC. You will be asked to correctly match this information from one list to that of corresponding information in another list.

EXAMPLE:

QUESTION: Compare the following electrical terms in column I with the appropriate hydraulic correlation in column II:

I	II
1. Voltage	A. GPM (gallons per minute)
2. Amperes	B. PSI (pounds per sq. inch)
3. Ohms	C. FL (friction loss)

ANSWER: 1 = B; voltage = PSI.
        2 = A; Amperes = GPM.
        3 = C; Ohms   = FL.

Entrance level Firefighter WRITTEN EXAMS are presented in a way so that they may determine each candidates current level of knowledge along with the candidates capabilities to grasp the functions required for entry level Firefighter.

EXAM SUGGESTIONS

Some suggestions to follow when taking the WRITTEN EXAM, include:
1. Be cooperative with exam proctor. His/her only function is to help you do your best on the exam.
2. Follow directions.
3. Make note of your exam test number and make sure that you have the correct number of pages in the exam booklet.
4. Read the whole question carefully; be sure that you know what the question asks and what the choices say.
5. Choose the answer that is generally correct; answer according to what is generally or usually true, not by what would be true in some particular case.
6. Use your time efficiently; The Firefighter written exam is not a speed test, but usually will not give you all the time that you might like to have. In any case, use all the time allowed. If the exam has a time limit, use either the entire time allowed or enough time to go through the entire test twice. Remember the difference between getting a job or not can be as close as one test question.
7. Make decisions; your decision should be one of the following:
   A. If you know the answer, answer this question now!
   B. That you can figure out the answer, but that it will take a lot of time, skip this question and come back to it later. Remember if you skip a question, you must also skip the same number on the answer sheet.
   C. That you do not know the answer and that you cannot figure it out, make a guess and answer this question now, unless you have been instructed not to do so. Remember, if you are not sure of the answer to a question and you decide to skip it and return to it later, your first instution is usually the best. Don't change the answer unless you are sure.
8. Don't give up; hang in there and give it a full effort.
9. Try not to change too many of your answers; remember that the best answer is the one that is usually or generally right, although if time permits review your exam.
10. Be at your best the day of the exam; be well rested, allow plenty of time to get to the exam, get there early. Relax the night before the exam.
11. Return your exam and exam materials to the exam proctor.

SECTION 2
GENERAL APTITUDE JUDGEMENT

SITUATION QUESTIONS

Entrance level Firefighter APTITUDE TEST are test in judgement. This category will represent normal Fire Department situations and require candidates to select a course of action that is accepted norm for the circumstances. EXAMPLES:

SITUATION
A friend of yours finds an official Firefighters badge and gives it to his younger brother to use as a toy.

JUDGEMENT OF SITUATION
The friends action is improper because the badge should be returned to the Fire Department.

SITUATION
You and a Firefighter friend of yours are on your way home from a late night movie. You both observe a fire in a hardware store with living quarters on the floors above. Your friend goes to warn the residents. You go to an alarm box a block away to report fire and then return to the fire scene.

JUDGEMENT OF SITUATION
Your action is improper because you should remain at the fire alarm box to direct the Fire Department to the fire scene.

SITUATION
An on duty Firefighter answers a Fire Department business phone line, but refuses to give the caller his name.

JUDGEMENT OF SITUATION
The Firefighters action is incorrect because Firefighters should give their name and rank as a matter of routine, when answering a departmental telephone.

SITUATION
While a Firefighter is conducting a routine fire inspection on a new business in the city, the owner ask the Firefighter a question concerning a problem that relates to another department within the city that the Firefighter has very little knowledge. The Firefighter suggest to the owner that he contact the appropriate department for his questions.

JUDGEMENT OF SITUATION
The Firefighters suggestion is the proper way to handle this situation.

SITUATION
During the extinguishment of a fire contained within a U.S. mail box a Firefighter used water to complete the extinguishment of the fire.

JUDGEMENT OF SITUATION
The use of water to extinguish a fire in this situation is improper because the water may damage the mail that is untouched by the fire, thus making the mail undeliverable.

SITUATION
During a fire that was preceded by an explosion, the automatic sprinkler system proved to be ineffective.

JUDGEMENT OF SITUATION
The reason for this is most likely because the explosion may have damaged the pipes that supply the sprinkler system.

SITUATION
When responding to alarms, many Fire Departments will use pre-established routes.

JUDGEMENT OF SITUATION
This is a good policy because:
1. Collision of responding apparatus is reduced.
2. Fastest routes are usually pre-planned.
3. Adverse road conditions may be avoided.

SITUATION
Public assembly buildings such as restaurants will have doors that open outwardly.

JUDGEMENT OF SITUATION
This is proper because in the event of a fire or other problem, it will prevent people that are exiting from panicking and jamming the doors in a closed position.

SITUATION
Engine company 82 usually conducts company inspections at irregular periods or intervals.

JUDGEMENT OF SITUATION
It is proper to conduct fire inspections at various times so that the inspection site will be seen in its normal condition and not in a "ready for inspection condition".

SITUATION
The company officer refused to turn the gas supply back on, after a house fire, for the home owner.

JUDGEMENT OF SITUATION
This is the proper procedure to follow because unburned gas may escape from open gas outlets/jets.

SITUATION
While responding to an alarm the fire officer was telling a personal story to the apparatus driver.

JUDGEMENT OF SITUATION
It would be improper to talk to the apparatus driver in this situation, other than to give orders or direction, because it could distract the apparatus driver from concentrating on driving.

SITUATION
During the training of recruit Firefighters, they will be trained in both engine company and truck company operations.

JUDGEMENT OF SITUATION
This is a proper procedure because at any fire they may be required to perform the duties of both the engine and ladder company.

SITUATION
During a fire the police department set up "fire lines" so as to keep unauthorized people out of the vicinity while the Fire Department is conducting the appropriate operations.

JUDGEMENT OF SITUATION
This procedure is appropriate since it will prevent hindrance with the Fire Departments operations.

SITUATION
While conducting a fire re-inspection a Firefighter observes the owner of the business being inspected correcting a violation during the re-inspection. The Firefighter informs the owner that he will come back at another time to complete the re-inspection.

JUDGEMENT OF SITUATION
The Firefighter used good judgement since it would be best to inspect at another time when the violation has been corrected completely.

SITUATION
At the scene of a fire a rookie Firefighter is given an order from his Captain that is inconsistent with the principles of firefighting that were taught at the fire academy. The Firefighters followed the orders.

JUDGEMENT OF SITUATION
This was the correct coarse of action for the Firefighter. He should always follow orders in an emergency situation (unless they endanger life and property) and then at a later time discuss the order that is questionable with his Captain.

SITUATION
At a house fire the Fire Captain orders is crew to open the doors and windows in a systematic way in order to ventilate the building.

JUDGEMENT OF SITUATION
This is a proper technique of ventilation so as to:
1. Increase visibility for the Firefighters
2. Reduce the toxic gases.
3. Control the travel of fire.

SITUATION
During a fire inspection a Firefighter suggest to the business owner that he remove large accumulations of combustible trash from the premises.

JUDGEMENT OF SITUATION
This advice is good since any source of ignition could cause the trash to catch fire.

SITUATION
During a fire inspection a Firefighter informs the of the business owner that he will have to remove high grass and weeds that are growing near the building.

JUDGEMENT OF SITUATION
This requirement is good because in the event of a fire the high grass and weeds could assist the travel of fire to the building and set it on fire.

SITUATION
During rescue operations involving the use of oxygen, a Firefighter request that family members in the location not smoke.

JUDGEMENT OF SITUATION
This is a proper request because the oxygen may cause the cigarette to flare-up dangerously.(O_2 supports combustion)

SITUATION
While fighting a hot and smoky fire the company officer orders his Firefighters to work in pairs.

JUDGEMENT OF SITUATION
This is a proper procedure resulting in safer operations since the Firefighters will be able to assist each other in the emergency operations.

SITUATION
The improper use of oxy-acetylene welders can create a fire hazard because of the failure to control or extinguish hot sparks.

JUDGEMENT OF SITUATION
This is an accurate statement.

SITUATION
Fire officers should make sure that their crews are aware that even a single 2 1/2" hose line with a 1" nozzle can deliver more than 2000 lbs of water per minute.

JUDGEMENT OF SITUATION
This is a true statement and important because of the possibility of building collapse.

SITUATION
Fire Department personnel should wear uniforms while on duty.

JUDGEMENT OF SITUATION
This is a true statement because other personnel and the public will better be able to recognize personnel that are on duty.

SITUATION
A Fire Captain ordered his nozzlemen to attack a fire involved in a group of evenly piled goods from above the fire.

JUDGEMENT OF SITUATION
This is the correct procedure because this is the position of the hose nozzle which will provide maximum water penetration to the goods exposed to the fire.

SITUATION
During company inspections the fire officer instructs his crew to inspect a building that is obviously made of fire resistant construction material such as bricks or natural stone.

JUDGEMENT OF SITUATION
This is the correct action to take because the interiors and contents of such a building are susceptible to fire.

SITUATION
While attempting to get as close to the seat of a very hot fire some Firefighters use a solid object such as a wall to shield themselves from the extreme heat, or they may even cool themselves by wetting down with small streams of water, or by spraying water into the heat waves that are coming from the fire.

JUDGEMENT OF SITUATION
These are all proper procedures to take so that the Firefighters may get as close to the seat of the fire as possible to allow them to direct their hose streams with accuracy.

SITUATION
During a pier fire the Fire Captain orders his crew to use sea water as the source of water to fight the fire.

JUDGEMENT OF SITUATION
This is a proper course of action since this will allow for an unlimited supply of water.

SITUATION
In attacking a fire on the fifth floor of a building, the Fire Captain orders his crew to advance the hose line up to the fourth floor prior to charging the hoseline with water.

JUDGEMENT OF SITUATION
This is the proper course of action since the hoseline will be easier to carry and handle prior to charging.

SITUATION
During a fire, in a warehouse, the commanding officer orders a hole in the roof for ventilation.

JUDGEMENT OF SITUATION
This a proper course of action since the fire will be fought more effectively by permitting the smoke and hot gases to escape.

SITUATION
Fire Department should make it a matter of policy to purchase the best apparatus and equipment and maintain them in the best condition at all times.

JUDGEMENT OF SITUATION
This is a the best policy since failure of apparatus and/or equipment in emergency situation may have serious consequences.

SITUATION
Fire Departments should have the policy that all Firefighters should wear their breathing apparatus in all structure fire situations.

JUDGEMENT OF SITUATION
This would be a good policy since the smoke from all fires are dangerous and may reduce the oxygen content of the air that is being breathed.

SITUATION
Prior to responding to an emergency, the Fire Captain orders his apparatus driver to remain in quarters until all of the crew are in their safety gear and sitting down in their seats with seat belts fastened.

JUDGEMENT OF SITUATION
This is a proper procedure to follow because it will reduce the possibility of injury to Firefighters.

SITUATION
While responding to a pier fire, the Fire captain orders the apparatus driver to avoid driving onto the pier.

JUDGEMENT OF SITUATION
This would be the proper procedure to follow since pier fires spread very rapidly and would endanger apparatus and crew.

SITUATION
When Fire Departments order new fire apparatus it is a good policy to order apparatus with enclosed cabs.

JUDGEMENT OF SITUATION
This is a good policy because it will protect Firefighters from the possible injury and adverse weather conditions.

SITUATION
While loading hose back on the apparatus, a Fire Captain instructs a rookie Firefighters to avoid bending the hose at places where it has been bent beforehand.

JUDGEMENT OF SITUATION
This is a good requirement since repetitive bending of fire hose in the same places will cause weakening of the hose in these locations.

SITUATION
During a fire in an unoccupied residence a rookie Firefighters finds a large sum of money in a closet and he gives the money to his Captain.

JUDGEMENT OF SITUATION
This would be the proper procedure for the rookie Firefighter or any Firefighter to follow.

SITUATION
As an off-duty Firefighter you observe fire apparatus in your rear view mirror, you then pull to the right of the street and stop.

JUDGEMENT OF SITUATION
This would be the proper course of action to take for off duty Firefighters or any citizen so as to allow the apparatus to pass safely.

SITUATION
Upon arrival to a fire scene a Firefighter was ordered by his Captain to take an axe from the apparatus and enter the building at the front door. The Firefighter then knocked down the door for entry.

JUDGEMENT OF THE SITUATION
This was not a proper procedure because the Firefighters should have checked to see if the front door was unlocked.

SITUATION
Most incident commanders will direct their crew to try to contain fires to their point of origin.

JUDGEMENT OF SITUATION
This is a good policy because by confining a fire to its area of origin property damage will be minimized.

SITUATION
A Fire Captain gives a rookie Firefighter a task to complete. The Firefighter starts the task and then encounters some difficulty in completing the task properly. The Firefighters continues to complete the task improperly.

JUDGEMENT OF SITUATION
This would be an improper procedure, the Firefighter should speak with the Fire Captain about his difficulties prior to completing the task.

SITUATION
A Fire Inspector instructed a school administrator to conduct the schools fire drills on the last Friday of every month just prior to the end of the school day.

JUDGEMENT OF SITUATION
This would be a unsatisfactory procedure to follow because fire drills should come at unexpected intervals and times.

SITUATION
While conducting a company fire inspection, the Fire Captain is told by the owner of the business that the inspection procedure is a waste of time and money. The Fire Captain tells the owner that he is just doing his job and continues the inspection.

JUDGEMENT OF SITUATION
This would be a poor way to handle this situation, the Fire Captain should explain to the owner the benefits of the inspection program.

SITUATION
While conducting search and rescue operations in a residence that is on fire, Firefighters usually pay particular attention to closets and spaces under beds and furniture.

JUDGEMENT OF SITUATION
These are proper procedures since victims, especially children, try to hide from danger in these places.

SITUATION
While conducting a company fire inspection the owner of the business ask the Fire Captain a technical question and the Fire Captain does not know the answer to the question. The Fire Captain informs the owner that he is not there to answer questions but to make an inspection.

JUDGEMENT OF SITUATION
This would be a poor response, the Fire Captain should advise the business owner that he does not know the answer but that he will research it and notify him.

SITUATION
As a rookie Firefighter you give an idea for improvement in station maintenance to a fellow Firefighter. At a later date this Firefighter is praised in front of you by your Captain for his excellent idea. You react by telling the Captain the whole story as to where the idea came from.

JUDGEMENT OF SITUATION
The proper way to handle this would be for you to do nothing about it, but next time make your suggestions to your Captain.

SITUATION
During a fire situation the Fire Captain orders his crew to place a hoseline directly in line with the travel of the fire.

JUDGEMENT OF SITUATION
This is a proper procedure since it will increase the chance of controlling the fire.

SITUATION
In fire situations, Fire Captains should make it a policy to use the smallest amount of water that is sufficient to put out the fire.

JUDGEMENT OF SITUATION
This is a good procedure to follow, mainly because it will reduce water damage.

SITUATION
While fighting a fire in a residence, a Firefighter calls a policeman to remove an individual that keeps trying to re-enter the residence to retrieve some important papers.

JUDGEMENT OF SITUATION
This is the proper course of action to take, because it is the Firefighters responsibility to protect life as well as property.

SITUATION
A Firefighters is opening windows in a smoked filled room in such a manner that he starts with the window nearest the entrance and follows the wall around the room until all the windows are open.

JUDGEMENT OF SITUATION
This is a good procedure because it will allow the Firefighter to find his way back to the entrance.

SITUATION
During a fire on the second floor of a building with central air-conditioning the incident commander orders a Firefighter to go to the basement and shut off the air-conditioning.

JUDGEMENT OF SITUATION
This would be a proper procedure since this would prevent the spread of smoke by the air conditioning system.

SITUATION
As a rookie Firefighter you are told by a fellow Firefighter that the basic reason for fire inspections is so that the Fire Department can exercise its authority on the public.

JUDGEMENT OF THE SITUATION
This is incorrect, the basic purpose of fire inspections is to obtain correction of conditions that create undue fire hazards.

SITUATION
During a fire in a multi-story building, your Fire Captain orders you to immediately go to the second floor of the building. The order was given prior to any knowledge as to what floor the fire is on.

JUDGEMENT OF SITUATION
This would be an improper course of action until it is determined it is safe to go to this floor.

SITUATION
A Firefighter is descending a flight of wooden stairs that have been charred by fire. The Firefighter descends backwards keeping close to the wall as he feels each step with his feet.

JUDGEMENT OF SITUATION
This is the proper procedure to use because the stairs are presumably better supported next to the wall.

SITUATION
In a smoke filled room with no visibility, a Firefighter tries to find a window by standing up and looking into the smoke.

JUDGEMENT OF SITUATION
This would be an unacceptable course of action, the Firefighter should get as low to the floor as he can, since this is where the coolest air will be found, and proceed to a wall and follow the wall until he comes to a window.

SITUATION
While fighting a fire in a smoke filled room with no visibility, the hoseline goes limp, no water, the Firefighters leave the hoseline to try and find an exit.

JUDGEMENT OF SITUATION
This would be a poor choice of action, the Firefighters should follow the hoseline back to the outside.

SITUATION
During a residential fire your Captain orders you to advance a hose line, but not to discharge water until you have actually located the fire.

JUDGEMENT OF SITUATION
This would be a proper course of action because it will be easier to maneuver the hoseline and it would reduce water damage.

SITUATION
After a Fire Captain and his crew has extinguished a house fire he is directed to make an effort to determine the origin of the fire.

JUDGEMENT OF SITUATION
This is a proper procedure because it may help to eliminate this cause of fire in the future.

SITUATION
Upon arrival at a fire in a tenement building the Fire Captain of the first in engine company orders his crew to advance a hoseline to the interior stairway of the building.

JUDGEMENT OF SITUATION
This would be a proper course of action because the stairway is a relatively safe and rapid means for evacuating tenants.

SITUATION
While fighting a fire in a warehouse your Captain notices that their are several places of fire origin located within the building, he orders his crew not to move anything after the fire is extinguished.

JUDGEMENT OF SITUATION
This would be a good procedure to follow, because several fires starting simultaneously is a strong indication of arson.

SECTION 3

READING COMPREHENSION VOCABULARY

EXPLANATION

Entrance Firefighter exams will have READING COMPREHENSION portions in order to test your ability to read, understand and retain information that you may be required to read during your performance as a firefighter.

You may be required to:
1. List facts or explain the meaning of words that you have read.
2. Note contradictions.
3. Interpret material that you have read.
4. Apply principles or opinions to the material that you have read.
5. Evaluate what you have read and to agree or differ with the point of view of the material.

Read the entire paragraph prior to looking at the questions and:
1. Sort out the principal view of the reading.
2. Locate distinctive components in the material.
3. Ascertain the significance of peculiar words that are used in the material.
4. Ascertain the distinctive method that the material is written so as to achieve its desired effect.
5. Use any prior knowledge that you have to complement the concept of the material.

READING QUESTIONS:

STATEMENT

Specific gravity equals the weight or mass of a given volume of a substance at a specified temperature as compared to that of an equal volume of another substance. Specific gravity applies to liquids only. Liquids of a specific gravity of less than 1 will float on water. The specific gravity of water is equal to 1 at 4 degrees Centigrade or 39 degrees Fahrenheit. Specific gravity is a ratio of weight to volume. When calculating the specific gravity of a substance, you are determining the ratio of the weight of a solid or liquid substance to the weight of an equal volume of water. The ratios are compared to water because it is simple to determine if the substance sinks or floats.

ACCORDING TO THE PREVIOUS STATEMENT

1. When comparing specific gravity:
A. It is compared to mercury.
B. It is compared to water.
C. It is compared to any substance of unequal ratios.
D. All of the above.

ANSWER = B

2. The specific gravity of water is equal to:
A. 1.
B. 5.
C. 10.
D. 15.

ANSWER = A

3. Specific gravity is a ratio of:
A. Length to height.
B. Weight to length.
C. Weight to volume.
D. Height to volume.

ANSWER = C

4. The term "specific gravity" applies to:
A. Gases.
B. Liquids.
C. Solids.
D. All of the above.

ANSWER = B

5. Liquids of a specific gravity of:
A. Less than 1 will float on water.
B. More than 1 will float on water.
C. Less than 1 will sink in water.
D. All of the above.

ANSWER = A

STATEMENT

Vapor density equals the weight of a vapor-air mixture resulting from the vaporization of a flammable liquid at equilibrium temperature and pressure conditions, as compared with the weight of an equal volume of air under the same conditions. Vapor density is the relative density of vapor gas, with no air present, compared to air. Vapor density is a ratio of gases. The vapor density of air is equal to 1, gases of less than 1 indicates that the gas is lighter than air. In order to compare vapor densities of gases, you must know that the molecular weight of air is equal to 29.

ACCORDING TO THE PREVIOUS STATEMENT

1. Vapor density of air is equal to:
A. 15.
B. 10.
C. 5.
D. 1.

ANSWER = D

2. The molecular weight of air is equal to:
A. 1.
B. 11.
C. 19.
D. 29.

ANSWER = D

3. Vapor density is equal to a ratio of:
A. Gases.
B. Liquids.
C. Solids.
D. All of the above.

ANSWER = A

4. Flammable liquid vapors have a vapor density:
A. Of less than 1 and will sink in the atmosphere.
B. Of greater than 1 and will sink in the atmosphere.
C. Of greater than 1 and will rise in the atmosphere.
D. None of the above.

ANSWER = B

5. Vapor density ratios are compared:
A. To air.
B. To solids.
C. To liquids.
D. In unlike conditions.

ANSWER = A

STATEMENT

There are two modes of which fire burns, the flaming mode and the smoldering mode. There are three stages of fire, the first stage of fire is a smoldering or incipient phase where the oxygen level is at 21% and the fire is burning at 100 degrees F., with a room temperature of 100 degrees F. The second stage is the flame producing phase where the oxygen level is at 21% to 15% with a fire and room temperature of 1300 degrees F. The third stage is a smoldering phase, or a decrease in heat generation, where the oxygen level is below 15% and the fire and room temperature is at 1000 degrees F.

ACCORDING TO THE PREVIOUS STATEMENT

1. How many modes are there that fire will burn?
A. There are two modes at which fire burns.
B. There are three modes at which fire burns.
C. There are four modes at which fire burns.
D. There are five modes at which fire burns.

ANSWER = A

2. How many stages of fire are there?
A. There are three stages of fire.
B. There are four stages of fire.
C. There are five stages of fire.
D. There are six stages of fire.

ANSWER = A

3. During the:
A. First stage of fire the room temperature = 1000 degrees F.
B. First stage of fire the room temperature = 100 degrees F.
C. Second stage of fire the room temperature = 100 degrees F.
D. Third stage of fire the room temperature = 100 degrees F.

ANSWER = B

4. During the:
A. First stage of fire the temperature of the fire = 100 degrees F.
B. Second stage of fire the temperature of the fire = 1000 degrees F.
C. Third phase of fire the temperature of the fire = 100 degrees F.
D. Third phase of fire the temperature of the fire = 1000 degrees F.

ANSWER = D

5. During the:
A. First phase of fire the oxygen content = 15%.
B. First phase of fire the oxygen content = below 15%.
C. Second phase of fire the oxygen content = 21% to 15%.
D. Third phase of fire the oxygen content = 21%.

ANSWER = C

STATEMENT

During the first stage of fire there is little or no decrease in the oxygen content of the interior atmosphere or in the average temperature of the interior atmosphere. Major damage will be caused by smoke during the first stage of fire. During the second stage of fire is when the major destruction will take place. During the third stage of fire there is a possibility of an inward rupture of window panes. If a Firefighters opens a door to an apartment and a smoke explosion occurs, the fire was most likely in the third stage of fire.

ACCORDING TO THE PREVIOUS STATEMENT

1. During the:
   A. First stage of fire there is a large increase in the oxygen content.
   B. First stage of fire there is a large increase in temperature.
   C. First stage of fire there is no decrease in oxygen content.
   D. None of the above.

   ANSWER = C

2. Major destruction:
   A. Takes place during the first stage of fire.
   B. Takes place during the second stage of fire.
   C. Takes place during the third stage of fire.
   D. Rarely takes place during any stage of a fire.

   ANSWER = B

3. During the:
   A. First stage of fire there is the possibility of the inward rupture of windows.
   B. Second stage of fire there is the possibility of the inward rupture of windows.
   C. Third stage of fire there is the possibility of the inward rupture of windows.
   D. All of the above.

   ANSWER = C

4. During the:
   A. First stage of fire there will be major damage by smoke.
   B. First stage of fire there will be no damage by smoke.
   C. Any stage of fire there will be no smoke damage created.
   D. Both B and C.

   ANSWER = A

5. If a firefighter opens the door to an apartment and an explosion occurs, the fire was most likely in the:
   A. First stage of fire.
   B. Second stage of fire.
   C. Third stage of fire.
   D. Between the first and second stage of fire.

   ANSWER = C

STATEMENT

Many people are not aware of the duties Firefighters perform, such as: training, fire prevention, and hydrant maintenance, etc. The general public usually are of the opinion that Firefighters are just sitting around the firehouse between fires. There are ways that Firefighters can assist in changing this image, such as: appearance, by greeting visitors who come to the firehouse, by their behavior on the street at a fire, and by treating the public in a courteous manner. Example: 75% of the rescues made by the average Fire Department take place at relatively small fires, not at dramatic large alarm fires. The public seldom hears about them because the Fire Departments seldom let the media know about Firefighters who have executed acts of heroism at routine fires.

ACCORDING TO THE PREVIOUS STATEMENT

1. When not fighting fires, Firefighters:
A. Lounge around the firehouse.
B. Work on public relation projects.
C. Repair tools and equipment.
D. None of the above.

ANSWER = D

2. The responsibility of improving the Fire Departments image should be placed on:
A. The media.
B. The public.
C. The people rescued by Firefighters.
D. The Fire Department.

ANSWER = D

3. Most rescues made by the Fire Department take place at:
A. Large alarm fires.
B. Special emergencies where no fire is involved.
C. Relative small fires.
D. Dramatic large fires.

ANSWER = C

4. The public rarely hears about rescues made by Firefighters at routine fires because:
A. Information about fires must be kept confidential.
B. Fire Departments seldom report these rescues to the media.
C. Most of these rescues take place late at night.
D. Reporters are not interested in covering routine fires.

ANSWER = B

5. What would be the best title for this material?
A. An Inside Look at the Fire Department.
B. Making the Most of Fire Prevention Week.
C. Improving the Fire Departments Public Image.
D. Brave Acts Performed by Firefighters.

ANSWER = C

STATEMENT

For many years the fire community acknowledged only three fire classifications. In 1960 the classifications were reorganized to show four fire classifications:
1. Class A fires = ordinary combustibles.
2. Class B fires = flammable liquids/gases.
3. Class C fires = electrical.
4. Class D fires = combustible metals.

CLASS "A" FIRES: these fires include ordinary combustibles such as wood, paper, fabric, solid plastics, and rubber. Class A fires normally involve fuels of an organic nature. These fires are the most common. Extinguishing agents for Class A type fires include water, some foam types, and multipurpose extinguishers.

CLASS "B" FIRES: these fires include all flammable and combustible liquids, gases, greases, and oils. One way to recognize a Class B fuel is by the container. No Class B fuel retains its own shape, because they are liquids and gases, These types of materials are usually stored in a strong rigid container. Extinguishing agents for Class B type fires include carbon dioxide, dry chemical, and foam types.

CLASS "C" FIRES: a Class C fire is one that involves energized electrical equipment. Very special importance must be given to the electrical non-conductivity of the extinguishing agent. Only when the electrical circuits have been de-energized may Class A and Class B extinguishing agents be used. Extinguishing agents suitable for Class C fires include dry chemical, carbon dioxide, and halon types.

CLASS "D" FIRES: when metal burns, they pose some very unique hazards. They burn extremely hot. They may actually react to ordinary extinguishing agents. Class D fires are fires involving such metals as sodium, magnesium, aluminum, uranium, and titanium. The hazards of a metal fire are so unique that ordinary extinguishing agents should generally not be used. Instead, extinguishing agents for Class D fires are those that have been specifically designed and approved for that type of application, such as dry powder types, Metal "X", etc.

ACCORDING TO THE PREVIOUS STATEMENT

1. There are how many classification of fire?
   A. 3.
   B. 4.
   C. 5.
   D. 6.

   ANSWER = B

2. Class A fires are fires involving:
   A. Ordinary combustibles.
   B. Flammable liquids/gases.
   C. Energized electricity.
   D. Combustible metals.

   ANSWER = A

3. Class B fires are fires involving:
   A. Ordinary combustibles.
   B. Flammable liquids/gases.
   C. Energized electricity.
   D. Combustible metals.

   ANSWER = B

4. Class C fires are fires involving:
   A. Ordinary combustibles.
   B. Flammable liquids/gases.
   C. Energized electricity.
   D. Combustible metals.

   ANSWER = C

5. Class D fires are fires involving:
   A. Ordinary combustibles.
   B. Flammable liquids/gases.
   C. Energized electricity.
   D. Combustible metals.

   ANSWER = D

6. Class A fires include combustibles such as all of the following, EXCEPT:
   A. Wood.
   B. Gases.
   C. Paper.
   D. Rubber.

   ANSWER = B

7. Class B fires include combustibles such as all of the following, EXCEPT:
   A. Combustible liquids.
   B. Rubber.
   C. Greases.
   D. Gases.

   ANSWER = B

8. Extinguishing agents for Class A type fires include all of the following, EXCEPT:
A. Water.
B. Some foam types.
C. Dry chemical.
D. CO2.

ANSWER = D

9. Extinguishing agents for Class B type fires include all of the following, EXCEPT:
A. Carbon dioxide.
B. Halon types.
C. Dry chemical.
D. All of the above.

ANSWER = B

10. Extinguishing agents for Class C type fires include all of the following, EXCEPT:
A. Dry chemical.
B. Carbon dioxide.
C. Halon types.
D. Dry powder.

ANSWER = D

11. Class C fires include:
A. Wood.
B. Gases.
C. Energized electricity.
D. Petroleum products.

ANSWER = C

12. Class D fires include combustible metals such as:
A. Sodium.
B. Magnesium.
C. Titanium.
D. All of the above.

ANSWER = D

STATEMENT

Fire Hose made of linen and unlined is basically tightly woven linen thread in the shape of a pipe. Because of the inherent properties of linen, soon after water has passed through a linen fire hose, the threads will expand causing the small spaces between them to close, making the hose watertight. Linen fire hose is inclined to deteriorate quickly if it is not completely dried after use or if located where it will be subjected to dampness or the climate. Linen fire hose is not normally constructed to withstand constant use or to be used where the material will be subjected to chafing from jagged or sharp areas.

ACCORDING TO THE PREVIOUS STATEMENT

1. Soon after water has passed through linen hose:
A. It will become water tight.
B. Small spaces will close.
C. Will continue to leak a large amount of water.
D. Both answers A and B are correct.

ANSWER = D

2. Unlined linen fire hose is best suited for use:
A. In an area where it will not be subjected to sharp objects.
B. For use at the fire academy.
C. As standard equipment on fire apparatus.
D. As a garden hose.

ANSWER = A

3. The LEAST appropriate use for unlined linen fire hose would be:
A. Emergency fire hose in a grocery store.
B. Emergency fire hose in an office building.
C. Emergency fire hose at a lumber yard.
D. Emergency fire hose in a warehouse.

ANSWER = C

4. Unlined linen fire hose is made of:
A. Plastic.
B. Rubber.
C. Synthetics.
D. Fabric.

ANSWER = D

5. Unlined linen fire hose is best if used:
A. Frequently.
B. Infrequently.
C. in rough terrain.
D. Daily.

ANSWER = B

STATEMENT

Fire hose has consistently been the most used piece of equipment in the Fire Service. Water is the extinguishing agent most frequently used in fighting fires. Fire hose is the means of moving water from the source to the area of need.

Fire hose is commonly accessible in the following sizes and is measured by internal diameter:
1. 3/4 inch.   5. 3 1/2 inch.
2. 1 inch.   6. 4 inch.
3. 1 1/2 inch.   7. 5 inch.
4. 2 1/2 inch.   8. 6 inch.

Fire hose is produced in many types of construction, based upon the expected field of service. The most conventional type of construction incorporates a rubber liner and one or two woven or molded outside covers. Some types of fire hose may be used for many different situations.

Usage of fire hose usually is in the category of suction hose or discharge hose.

ACCORDING TO THE PREVIOUS STATEMENT

1. The most used piece of equipment in the Fire Service is:
   A. Water.
   B. Ladders.
   C. Fire hose.
   D. Fire extinguisher.

   ANSWER = C

2. The most common extinguishing agent used in fighting fires is:
   A. Water.
   B. Foam.
   C. Dry chemical.
   D. Fire hose.

   ANSWER = A

3. Fire hose is generally available in all of the following sizes, EXCEPT:
   A. 4 inch.
   B. 5 inch.
   C. 6 inch.
   D. 8 inch.

   ANSWER = D

4. The most common type of fire hose construction is:
A. Dacron.
B. Unlined.
C. Rubber lined with two woven covers.
D. Rubber lined with one woven cover.

ANSWER = C

5. Usage of fire hose can be divided into two types, which are:
A. Pressure and non-pressure.
B. Suction and discharge.
C. Suction and non-pressure.
D. Discharge and pressure.

ANSWER = B

STATEMENT

Fire hose is designed for the use of transporting water and should never be utilized for anything other than that function. The demand for routine testing and maintenance is evident. A burst or ruptured hose at a fire scene may be life threatening. Not only does this stop the water supply, but it creates a significant hazard to those in the area because of the flailing action of the hose that is out of control. In order to help insure that fire hose will meet performance standards, fire hose should be tested at least annually.

The most commonly used fire hose is the woven-jacket type, with the possible exception of booster and hard suction hoses. Fire hose should be loaded on apparatus so as to allow air to circulate freely, although this is not always feasible. High humidity climates make this a more complex problem to deal with. Fire hose should be removed from apparatus and reloaded, at certain time intervals, in a different order so that the hose that was on the bottom will be in a different position and all the hose should be in a different order, so that the kinks may be eliminated. When hose is subjected to rain or snow, it should be placed on wooden racks with an air space below. The hose bed and all of the exposed fire hose should be covered with some sort of a waterproof cover. A waterproof cover not only protects the fire hose from moisture, they also will reduce deterioration of the fire hose from continued exposure to the sun. Fire hose should be thoroughly cleaned and dried periodically.

33

During emergency incidents, fire hose should be positioned without sharp kinks that can cause extreme internal stress. The proper positioning of fire hose should include avoiding sharp or rough objects. Except in an emergency situation, vehicles should never drive directly over fire hose. A vehicle may drive over fire hose with the use of a hose bridge, which will keep the weight of the vehicle off of the fire hose, thus the vehicle will not cause any damage to the fire hose. Also hose belts or rope tools should be used whenever fire hose is hoisted up ladders or other elevations, so as to take the excessive stress off of the couplings.

Firefighters must be aware of the damage that water hammer can cause to fire hose. Water hammer can take place when a valve or nozzle is closed too quickly. The water will continue to move within the fire hose which will cause increased pressure. This increased pressure could damage the fire pump or cause the fire hose to rupture. Therefore, firefighters should always close valves and nozzles slowly.

To protect the discharge side of the pump, pressure surges are averted by the use of relief valves and speed governors in union with pumping operations. During relay or tandem pumping this is particularly important. Relay and tandem pumping is needed when supplying water through fire hose in lengths of over 2000 feet or when delivering water up a grade. During this procedure an pumper is placed at the water source and delivers water to another pumper or to the fire, depending on the length of the hose lay. Relay or tandem pumping is needed because of the added friction loss created by the fire hose over a long distance.

Special care should be taken, during cold climate conditions, to prevent water from freezing inside the fire hose. After the flow of water is initiated, a constant flow of water should continue until the fire hose is not needed and is going to be removed. Avoid sharp bends when ice has developed. Ice inside fire hose may act like a knife. It may be necessary to chop frozen fire hose out of the ice. Care must used when doing this so as not to damage hose.

Firefighters should avoid placing fire hose too close to the fire, because the fire hose may be scorched or burned. In wildland fire conditions, the use of unlined fire hose is a benefit because water will ooze through the fabric and obstruct the material from burning by keeping the fabric wet.

After fire incidents, fire hose should be laid out, at the fire station, and brushed cleaned with a mild soap and water to remove dirt, etc. The use of a small hose will be useful during the cleaning procedure. The fire hose must be thoroughly dried after it is cleaned. Fire hose may be dried by hanging it in a hose tower, air drying on horizontal racks, or in a heated forced air drying machine. Dry fire hose should be rolled and placed on racks. Rubber fire hose, such as booster hose only needs to have dirt, etc. wiped off. Many of the larger Fire Departments have a central fire hose cleaning and maintenance facility.

As far as fire hose is concerned, good record keeping is vital along with the normal use, maintenance, and storage. Each length of fire hose should be identified with a number. When hose changes take place, the lengths and order of fire hose on equipment should be recorded. This will allow firefighters to calculate the length of hose leads by looking at the inventory number on the fire hose and comparing it with the inventory list of the apparatus. Fire hose certification test should also be recorded and periodic replacement may be sanctioned.

ACCORDING TO THE PREVIOUS STATEMENT

1. Fire hose is designed for the use of:
A. Adding pressure to the water.
B. Reducing water pressure.
C. Moving water.
D. Holding water.

ANSWER = C

2. Fire hose should be tested at least:
A. Twice a year.
B. Once a year.
C. Once every two years.
D. Does not need to be tested.

ANSWER = B

3. It is best to load fire hose:
A. So that air will circulate.
B. Tightly.
C. Always in the same position.
D. Always in the same order.

ANSWER = A

4. Fire hose should be:
A. Covered in the hose bed.
B. Cleaned and thoroughly dried.
C. Protected from the sun.
D. All of the above.

ANSWER = D

5. As far as vehicles driving over fire hose, which of the following is true:
A. Should never happen.
B. Will not cause any damage.
C. Is alright only during drills.
D. Should never happen, except in an emergency.

ANSWER = D

6. The condition that occurs when a valve or nozzle is abruptly closed, while water continues to move in the hose, causing increased pressure and may damage or rupture the hose is called:
A. Head pressure.
B. Back pressure.
C. Water Hammer.
D. Water damage.

ANSWER = C

7. To prevent pressure surges during pumping operations, relief valves and engine speed governors are operated:
A. In union with pumping operations.
B. Opposite of each other.
C. At all times.
D. None of the above.

ANSWER = A

8. Tandem or relay pumping is necessary when supplying water through hose lengths of over:
A. 1000 feet.
B. 1200 feet.
C. 1500 feet.
D. 2000 feet.

ANSWER = D

9. Relay pumping is basically necessary because:
A. Of Head pressure.
B. Of Back pressure.
C. Of Friction.
D. Nozzle reaction.

ANSWER = C

10. To prevent water from freezing inside fire hose in cold climate conditions:(water flowing)
A. Shut off hose line when it is not being used.
B. Continue flowing water even when hose line is not being used.
C. Use increased pressures.
D. Used decreased pressures.

ANSWER = B

11. In wildland fires, what helps prevent unlined fire hose from being burned?
    A. Water leaking through the fabric.
    B. The added insulation.
    C. Water accumulation in the area.
    D. None of the above.

    ANSWER = A

12. After fire incidents, fire hose should be:
    A. Rinsed with clear water.
    B. Rinsed with mild soap.
    C. Brushed and dried.
    D. All of the above.

    ANSWER = D

13. Rubber hose, such as booster hose, should be:
    A. Washed and dried.
    B. Dried only.
    C. wiped clean.
    D. None of the above.

    ANSWER = C

14. In addition to normal usage, cleaning, and storage of fire hose, it is necessary to keep:
    A. Records of fire hose.
    B. Old fire hose.
    C. Old couplings.
    D. None of the above.

    ANSWER = A

STATEMENT

In the fire service, ventilation is the method of opening up a building or structure that is involved in fire so as to liberate the accumulated smoke, gases, and heat. If Firefighters do not have the required knowledge of the principles of ventilation, it could create unnecessary damage, and or injuries. Ventilation by itself will not extinguish fires, but when used in an prudent manner it will allow Firefighters to locate the fire easier and faster with less difficulty and risk.

ACCORDING TO THE PREVIOUS STATEMENT

1. The major result of failing to apply the principles of ventilation at a fire may be:
    A. Injury to the Firefighters.
    B. Inappropriate use of equipment and apparatus.
    C. Waste of water.
    D. Disciplinary action.

    ANSWER = A

2. The best reason for ventilation is that it:
A. Will lower the temperature of the fire.
B. Eliminates the need for breathing apparatus.
C. Allows the Firefighters to advance closer to the fire.
D. Reduces smoke damage.

ANSWER = C

3. The use of ventilation principles would be the LEAST useful in a fire involving:
A. Apartment building.
B. Grocery market.
C. Bank.
D. Lumber yard.

ANSWER = D

4. For a well trained and well equipped Firefighter, ventilation is:
A. Not needed.
B. Unimportant.
C. A basic procedure.
D. Rarely used.

ANSWER = C

5. Ventilation of a fire building or structure will release:
A. Excessive water.
B. Occupants.
C. Smoke and heat, but not the toxic gasses.
D. Smoke, heat, and the toxic gases.

ANSWER = D

STATEMENT

Within the Fire Service, the Hurst tool and shears have many uses in both rescue and forcible entry. Each tool, the "jaws of life" and the shears, may be used individually or simultaneously from the same power supply. The power supply produces hydraulic oil pressure that circulates through flexible lines. These lines must be attached to the tools before they will function. The fluid used for the tools is not a standard hydraulic oil, it is an oil of an acid base that can cause burns to the skin, if contacted. Each tool weighs about 60 pounds. The power unit weighs about 40 pounds. The flexible arms of the tool are about 2 1/2 feet long and made of forged titanium. The tensile strength of the arms is 155,000 pounds. The actual force exerted at the tip of the "jaws" is 10,000 pounds in both the push or pull operation. The "jaws of life" are used for prying and separating, without the ability to perform cutting operations.

To operate the tool firefighters must grasp the back portion with both hands, using the thumb on the right hand to control the opening and closing of the jaws. The control must also be operated by left handed firefighters with their right thumb.
Firefighters may cut the post, which anchors a car top to the body of the car, with one cutting motion, with the use of the "shears". When this is completed firefighters may extricate victims from the car by literally peeling the car top back. Firefighters should always be aware of sharp and jagged edges while operating the shears. Using these tools, firefighters can accomplish, in seconds, cutting operations which include windshield post and metal-clad door frames. Firefighters, with the added use of chains, can pull a steering wheel and column completely out of a car. The tool is capable of lifting the weight of a car or a truck.

ACCORDING TO THE PREVIOUS STATEMENT

1. The hurst tool and shears:
   A. Must be operated separately.
   B. Must be operated together.
   C. May be operated separately or together from one power supply.
   D. If operated together, must have two power supplies.

   ANSWER = C

2. The power supply for the hurst tool will generate:
   A. Vacuum pressure.
   B. Air pressure.
   C. Hydraulic oil pressure.
   D. Water pressure.

   ANSWER = C

3. The hurst tool weighs about:
   A. 40 pounds.
   B. 50 pounds.
   C. 60 pounds.
   D. 70 pounds.

   ANSWER = C

4. The jaws of the hurst tool are flexible and made of forged:
   A. Aluminum.
   B. Steel.
   C. Magnesium.
   D. Titanium.

   ANSWER = D

5. The length of the jaws on the hurst tool are about:
A. 1 1/2 feet long.
B. 2 feet long.
C. 2 1/2 feet long.
D. 36 inches long.

ANSWER = C

6. The tensile strength of the jaws are:
A. 55,000 pounds.
B. 75,000 pounds.
C. 100,000 pounds.
D. 155,000 pounds.

ANSWER = D

7. The actual force exerted at the tip of the jaws is:
A. 5,000 pounds push and 10,000 pounds pull.
B. 5,000 pounds pull and 10,000 pounds push.
C. 10,000 pounds push and 10,000 pounds pull.
D. 5,000 pounds push and 5,000 pounds pull.

ANSWER = C

8. To control the opening and closing of the jaws, the firefighter must use:
A. Right hand thumb.
B. Left hand thumb.
C. Right hand or left hand thumb.
D. Left hand thumb, if the firefighter is left handed.

ANSWER = A

9. When cutting operations are necessary:
A. Increase the power of the tool.
B. Decrease the power of the tool.
C. Use the shears.
D. Both A and B.

ANSWER = C

10. With the use of the hurst power and chains, it is possible to:
A. Pull a steering wheel completely off.
B. Pull a steering column completely off.
C. Lift the weight of a car or a truck.
D. All of the above.

ANSWER = D

STATEMENT

Spontaneous ignition is the ignition due to chemical reaction or bacterial action in which there is a slow oxidation of organic compounds until the material ignites. Usually there is sufficient air for oxidation but not enough ventilation to carry heat away as it is generated. Substances that are susceptible to spontaneous ignition are able to catch fire without an external source of heat. In all cases of spontaneous ignition, heat of oxidation must be produced more rapidly than it is dispersed.

ACCORDING TO THE PREVIOUS STATEMENT

1. Spontaneous ignition is due to:
A. Chemical reaction.
B. Bacterial action.
C. Both A and B.
D. None of the above.

ANSWER = C

2. Spontaneous ignition requires that oxidation:
A. Take place without air.
B. Take place with enough air to carry heat away as it is generated.
C. Take place with enough air for the oxidation, but not enough to carry the heat away as it is generated.
D. None of the above.

ANSWER = C

3. In all cases of spontaneous ignition the heat of oxidation must be:
A. Dispersed more rapidly than it is produced.
B. Produced more rapidly than it is dispersed.
C. Not produced at all.
D. Not dispersed at all.

ANSWER = B

5. Substances that are susceptible to spontaneous ignition are able to catch fire:
A. Without an external source of heat.
B. Only with an external source of heat.
C. Under any conditions.
D. None of the above.

ANSWER = A

STATEMENT

Size-up is the mental evaluation made by the fire officer in charge., which enables him to determine a course of action. Size-up is a form of reconnaissance. Size-up includes such factors as time, exposure, property involved, nature and extent of the fire, available water supply and other fire fighting facilities. The size-up report is usually given via radio. In order for a fire officer to make a correct size-up, he should maintain a disciplined mind, and be instructed to think logically during confusion and excitement.

The four stages of size-up include: anticipating the situation, gathering the facts, evaluating the facts, and determining the procedures. Some things to consider at the time of size-up include: the location of the fire, location of the fire within the structure, size of fire, smoke and gases that are being generated, type of contents within the structure, potential of danger to Firefighters, and potential of danger to occupants.

ACCORDING TO THE PREVIOUS STATEMENT

1. Size-up enables the fire officer to determine:
A. What caused the fire.
B. Where the fire started.
C. Course of action.
D. What is burning.

ANSWER = C

2. When a fire officer is making a size-up he should consider:
A. Time of day.
B. Exposures
C. Property involved.
D. All of the above.

ANSWER = D

3. Size-up is usually given via:
A. Telephone.
B. Radio.
C. Face to face.
D. Headquarters.

ANSWER = B

4. All but one of the following are considered stages of size-up, which one is not:
A. Anticipating the situation.
B. Gathering the facts.
C. Evaluating the facts.
D. Performing the proper procedures.

ANSWER = D

42

5. All but one of the following should be considered at the time of size-up, which one is not:
A. Location of fire.
B. Size of fire.
C. Who started the fire.
D. Smoke and gases being generated.

ANSWER = C

STATEMENT

Foam fire extinguisher use three chemicals, bicarbonate of soda, aluminum sulphate and a stabilizer. The small interior section will usually comprise a 50% water solution of aluminum sulphate. The larger outer section will comprise of approximately 7% bicarbonate of soda, 3% stabilizer, and 90% water. The stabilizer is used to make the bubbles smaller in diameter and more persistent. When the fire extinguisher is inverted the chemicals of the two sections will intermix, which creates the foam and causes it to expel. The main extinguishing agent is the small bubbles of carbon dioxide gas that will be trapped within the walls of the aluminum hydrate, which forms a hardy, durable, flexible and sticky foam that is able to withstanding a considerable amount of abuse. 90% of the foam consist of carbon dioxide gas volume, although approximately 75% of the foam is water by weight. The foam will create a blanket of bubbles over the ignited substance, which will eliminate the air and cool the surface. Cellulose products such as fabric are made fire resistant by the strengthened bubbles containing aluminum hydrate. Both horizontal and vertical surfaces are coated and isolated from the heat by the foam adhering wherever it is utilized including floating over liquids. The foam is harmless to Firefighters. The foam creates less of a wetting effect than water. The extinguisher is not effective on fires involving alcohol. The extinguisher should be protected from low temperatures.

ACCORDING TO THE PREVIOUS STATEMENT

1. The stabilizer in a foam fire extinguisher, has the primary function of preventing:
A. Clinging of the foam to the extinguisher.
B. Early chemical reaction of the ingredients.
C. Swift breaking-up of the carbon dioxide bubbles.
D. Vaporization of the aluminum sulphate.

ANSWER = C

2. The primary justification for removing the air from an ignited substance with the use of a layer of foam is that:
A. Fire needs air to continue burning.
B. The layer of foam will allow the heat to escape.
C. The ignition temperature of the substance will be lowered by the foam.
D. Heat from the fire will break down the carbon dioxide.

ANSWER = A

3. The reason that foam is 90% carbon dioxide by volume and 85% water by weight is that:
A. Carbon dioxide is a pure gas.
B. Carbon dioxide weighs more than air.
C. Carbon dioxide occupies less volume than water.
D. Carbon dioxide has less density than water.

ANSWER = D

4. Foam fire extinguisher are not effective on fires involving:
A. Ignited wood.
B. Ignited fabric.
C. Ignited gasoline.
D. Ignited alcohol.

ANSWER = D

5. Foam fire extinguisher expel their ingredients by the use of:
A. The force created by the stabilizer acting in the solution.
B. A Firefighters operating the hand-pump.
C. The gas pressure created by the chemical reaction.
D. The expanding of the water in the inside section.

ANSWER = C

STATEMENT

Selection of the best portable fire extinguisher for a given situation depends on:
1. The nature of the combustibles which might be ignited.
2. The potential severity (size, intensity, and speed of travel) of any resulting fire.
3. The effectiveness of the extinguisher on that hazard.

4. The ease of use of the extinguisher.
5. The personnel available to operate the extinguisher and their experience and training along with emotional reactions as influenced by their training.
6. The ambient temperature conditions and other special atmospheric considerations (wind, draft, presence of fumes).
7. The suitability of the extinguisher for its environment.
8. Any anticipated adverse chemical reactions between the extinguishing agent and the burning material.
9. Any health and operational safety concerns (exposures of operators during the fire control)
10. The upkeep and maintenance requirements for the extinguisher.

Portable fire extinguisher are designed to cope with fires of limited size and are necessary and desirable even though the property may be equipped with automatic sprinkler protection, standpipe and hose systems, or other fixed fire protective equipment.

The initial selection of the type and capacity of an extinguisher is based on the hazards of the area to be protected. NFPA #10, Standard for the Installation, Maintenance, and use of Portable Fire Extinguisher, i.e. "NFPA Extinguisher Standard", has established three hazard levels in order to provide a simplified method of determining the portable size of a fire relative to the kind of incipient fire and its potential severity:

Light Hazard:
Where the amount of combustibles or flammable liquids present is such that fires of small size may be expected. These may include offices, schoolrooms, churches, assembly halls, telephone exchanges, etc.

Ordinary Hazards:
Where the amount of combustibles or flammable liquids present is such that fires of moderate size may be expected. These may include mercantile storage and display areas, auto showrooms, parking garages, light manufacturing areas, warehouses not classified as extra hazard, school shop areas, etc.

Extra Hazards:
Where the amount of combustibles or flammable liquids present is such that fires of severe magnitude may be expected. These may include woodworking areas, auto repair shops, aircraft servicing areas, warehouses with high-piled combustibles (over 15 feet in solid piles, over 12 feet in piles that contain horizontal channels), and areas involved with processes such as flammable liquid handling, painting, dipping, etc.

The class of hazard can influence the type of extinguisher selected as well as the size of fire extinguishing capability (i.e., 2 1/2 gallon capacity stored pressure or pump tank water extinguisher are rated 2-A and are only suitable for light or ordinary hazard protection. When extra hazard conditions exist, multipurpose dry chemical extinguisher having ratings of 3-A to 40-A will provide the degree of protection needed).

ACCORDING TO THE PREVIOUS STATEMENT

1. Selection of the best portable fire extinguisher for a given situation depends on:
A. Nature of combustibles that are ignited.
B. Potential severity of any resulting fire.
C. Ease of use.
D. All of the above.

ANSWER = D

2. Portable fire extinguisher are designed to cope with fires:
A. Of limited size.
B. Of any size.
C. In outside areas only.
D. In inside areas only.

ANSWER = A

3. The initial selection of the type and capacity of an extinguisher is based upon:
A. The size of the fire.
B. The location of the fire.
C. The hazard of area to be protected.
D. The hazard of the contents of extinguisher.

ANSWER = C

4. The NFPA Standard for the Installation, Maintenance and use of Portable Fire Extinguisher is:
A. NFPA #8.
B. NFPA #9.
C. NFPA #10.
D. NFPA #15.

ANSWER = C

5. Light Hazard occupancies include all of the following, EXCEPT:
A. Offices.
B. Schoolrooms.
C. Hospitals.
D. Assembly halls.

ANSWER = C

6. Ordinary Hazard occupancies include all of the following, EXCEPT:
A. Mercantile storage.
B. Auto repair shops.
C. Auto showrooms.
D. Parking garages.

ANSWER = B

7. Extra Hazard occupancies include all of the following, EXCEPT:
A. Aircraft servicing areas.
B. Auto painting shops.
C. Warehouses with high-piled combustibles.
D. School auditoriums.

ANSWER = D

8. When Extra Hazard conditions exist, it is recommended that: which of the following ratings be provided?
A. 2 1/2 gallon water.
B. Dry chemical, 2-A to 15-A.
C. Dry chemical, 3-A to 40-A.
D. Multipurpose dry chemical, 3-A to 40-A.

ANSWER = D

STATEMENT

In general, liquefied gas extinguisher: bromotrifluoromethane (halon 1301) and bromochlorodifluoromethane (halon 1211), have features and characteristics similar to CO-2 extinguisher. The bromotrifluromethane (halon 1301) extinguisher has never been available in a size larger than 2 1/2 lbs. It has a listed rating of 2-B:C which is below the minimum requirements. This extinguisher does not appear to have a large advantage over other liquified gas extinguisher.

The bromochlorodifluromethane (halon 1211) extinguisher is available in a wide range of sizes, with listed ratings of 2-B:C to 10-B:C. The agent is similar to CO-2 in that it is suitable for cold weather installation, is noncorrosive, and leaves no residue. It is considerably more effective on small Class A fires than CO-2; however water may still be needed as a follow-up to extinguish glowing embers and deep-seeded burning. Compared to CO-2 on a weight-of-agent basis, bromochlorodifluoromethane (halon 1211) is at least twice as effective. When discharged, the agent is in the combined form of a gas/mist with about twice the range of CO-2. To some extent, windy conditions or strong air currents may make extinguishment difficult by causing the rapid dispersal of the agent. The shell for the halon 1211 extinguisher are light weight aluminum or mild-steel and weigh considerably less than CO-2 cylinders.

ACCORDING TO THE PREVIOUS STATEMENT

1. Liquified gas extinguisher such as halon 1301 and halon 1211 have features and characteristic similar to:
A. CO-2 extinguisher.
B. 2-A 10 BC extinguisher.
C. Dry chemical extinguisher.
D. Multipurpose dry chemical extinguisher.

ANSWER = A

2. Halon 1211 extinguisher are not available in sizes above:
A. 2 1/2 lbs.
B. 3 lbs.
C. 5 lbs.
D. None of the above.

ANSWER = D

3. Halon 1301 extinguisher are not available in sizes above:
A. 2 1/2 lbs.
B. 3 lbs.
C. 5 lbs.
D. None of the above.

ANSWER = A

4. Halon 1211 is:
A. Not suitable for cold weather installation.
B. Not similar to CO-2.
C. Noncorrosive.
D. The one that leaves a residue.

ANSWER = C

5. Halon 1211 as compared to CO-2 on a weight-to-agent basis, is:
A. About 1/2 as effective.
B. About 2 times as effective.
C. Has about 1/2 the range.
D. Uses a heavier container.

ANSWER = B

STATEMENT

Fire strategy is the plan of attack on a fire. Fire strategy should make prime use of equipment and personnel and take into consideration fire behavior, the nature of the occupancy, environmental conditions, and weather factors. Fire tactics are the various maneuvers that can be employed in a strategy to successfully fight a fire. Even though no two fires are alike, it is possible to lay down general plans for firefighting operations primarily because the elements of similarity are sufficient enough to establish fire strategy and fire tactics that are applicable in a variety of situations. In any fire remember for covering all points of a fire, cover the front and rear, and over and under along with covering all exposures.

ACCORDING TO THE PREVIOUS STATEMENT

1. Fire strategy is:
A. The attack on a fire.
B. The evaluation of a fire scene.
C. The plan of attack on a fire.
D. The critique of a fire.

ANSWER = C

2. Fire tactics are:
A. The extinguishment of a fire.
B. The critique of a fire.
C. The various apparatus and equipment that may be employed in a strategy to successfully fight a fire.
D. The various maneuvers that can be employed in a strategy to successfully fight a fire.

ANSWER = D

3. Fire strategy should:
A. Make prime use of equipment and personnel.
B. Take fire behavior into consideration.
C. Take the weather into consideration.
D. All of the above.

ANSWER = D

4. It is possible to lay down general plans for firefighting, because:
A. No two fires are alike.
B. All fires are alike.
C. The elements of similarity are sufficient enough to establish tactics and strategy.
D. The elements of tactics and strategy are always exactly the same.

ANSWER = C

5. In covering all points of a fire officers should remember to cover:
   A. The front and rear.
   B. Over and under.
   C. All exposures.
   D. All of the above.

   ANSWER = D

STATEMENT

When a superior officer takes over the command of a large fire from a subordinate officer that has been in charge for some time, he should make a review the actions taken up to this time. The first action that an officer should check when relieving another officer at a fire scene, is whether the fire is adequately surrounded by the apparatus and Firefighters available. If the structure is fully involved in fire, which is free of exposures on all sides, it is normally best to direct water streams to the structure from three sides rather than from all four sides, mainly because the heat, smoke and gases will be driven away from the Firefighters, permitting them to work more proficiently. Remember that the most common mistake that the first in officer may have made is not to have requested the proper additional help needed, because manpower is the most critical at the early stages of a fire.

ACCORDING TO THE PREVIOUS STATEMENT

1. When a superior officer takes over a fire from a subordinate officer:
   A. Make a check of the actions taken up to this time.
   B. Cancel all previous orders.
   C. Call for additional help.
   D. Make a list of previous improper actions.

   ANSWER = A

2. When an officer relieves another officer at a fire, his FIRST action taken should be:
   A. Check for adequate personnel and apparatus.
   B. Call the media.
   C. Check to see if the fire is adequately surrounded by apparatus and Firefighters.
   D. See if the fire was deliberately started.

   ANSWER = C

3. When a structure is fully involved in fire, and is free on all sides, it is best to direct water streams to the structure from:
A. The front.
B. The rear.
C. Three sides.
D. All four sides.

ANSWER = C

4. The most common mistake that the first in officer will make is:
A. Deliver too much water.
B. Neglect to call for additional help.
C. Request too much help.
D. Attack the fire too aggressively.

ANSWER = B

5. Manpower: most critical at what stage of fire?
A. Early stages.
B. Middle stages.
C. Late stages.
D. None of the above.

ANSWER = A

STATEMENT

Centrifugal pumps employ a certain principle of force in pumping. The power or force to create pressure is exerted from the center. The revolving motion of the impeller will whirl water introduced at the center toward the outer edge of the impeller. Here it is trapped by the pump casing and is forced to the discharge outlet. In the centrifugal pump the power is transmitted from the drive shaft, through the pump transmission, and intermediate gear to the impeller shaft. With the quantity remaining constant in a centrifugal pump the pressure will increase at a rate equal to the square of the speed increase. The pump speed is greater than the engine speed in a centrifugal pump. Centrifugal pumps cannot create a vacuum.

ACCORDING TO THE PREVIOUS STATEMENT

1. The force or power in centrifugal pumps is exerted from:
A. The outside edge of the of the impeller.
B. The center of the pump.
C. The discharge outlet.
D. The intake valve.

ANSWER = B

2. Centrifugal pumps power is transmitted FROM:
A. The drive shaft.
B. The pump transmission.
C. The intermediate gear.
D. The impeller.

ANSWER = A

3. In centrifugal pumps the water is introduced at:
A. Outer edge of the impeller.
B. Center towards the outer edge.
C. Outer edge towards the center.
D. The pump casing.

ANSWER = B

4. Centrifugal pumps are NOT capable of creating:
A. Pressure.
B. Force.
C. Vacuum.
D. Discharge.

ANSWER = C

5. In centrifugal pumps, with the quantity remaining constant, the pressure will increase at a rate equal to:
A. The square of the quantity increase.
B. The square of the volume increase.
C. The square of the discharge increase.
D. The square of the speed increase.

ANSWER = D

STATEMENT

Water under pressure, in the volute of a centrifugal pump is prevented from returning to the suction side of the pump by the close fit of the impeller hub to a stationary wear ring at the eye of the impeller, and hydraulic pressure due to the velocity created by the centrifugal force. The volute enables the centrifugal pump to handle the increasing quantity of water towards the discharging outlet and at the same time permit the velocity of the water to remain constant or to decrease gradually maintaining the continuity of flow. The volute is a progressively expanding waterway which converts velocity to pressure as the velocity remains constant. The volute principle is the design of the water passageway in centrifugal pumps.

ACCORDING TO THE PREVIOUS STATEMENT

1. In the volute of a centrifugal pump, the water under pressure is prevented from returning to the suction side of the pump by:
   A. Close fit of the impeller.
   B. Hydraulic pressure created in the pump.
   C. Both A and B.
   D. None of the above.

   ANSWER = C

2. Centrifugal pumps volute enables the pump to:
   A. Handle the increasing quantity of water towards the discharge outlet.
   B. Permit the velocity of the water to remain constant.
   C. Permit the velocity of the water to decrease.
   D. All of the above.

   ANSWER = D

3. The volute of a centrifugal pump converts:
   A. Pressure to velocity.
   B. Velocity to pressure.
   C. Pressure to volume.
   D. Volume to pressure.

   ANSWER = B

4. The volute principle in a centrifugal pump is the design of:
   A. The airway.
   B. The impeller.
   C. Discharge valve.
   D. The water passageway.

   ANSWER = D

5. The volute in a centrifugal pump is a:
   A. Decreasing waterway.
   B. Expanding waterway.
   C. Constant waterway.
   D. Unchanging waterway.

   ANSWER = B

STATEMENT

The present recognized capacities of pumps for Fire Department pumpers are: 500, 700, 1000, 1250, and 1500 gallons per minute (GPM), although some larger capacity pumpers have been built. In order for a Fire Department pumper to meet standard requirements, it must deliver its rated GPM capacity at 150 pounds per square inch (PSI). The pumper further must deliver 70% of its rated capacity at 200 PSI and, 50% of its rated capacity at 250 PSI net pump pressure. A Fire Department pumper must be provided with adequate inlet and discharge pump connections, pump and engine controls, gauges, and other instruments.

Some Fire Department pumpers are referred to as a "Triple Combination Pumper". To qualify as a triple combination pumper, a pumper must have a fire pump, a hose compartment, and a water tank. In addition, a Fire Department pumper will usually carry ladders, tools and equipment, hose, and other accessories. The location of the three components of a triple combination pumper may vary with each manufacture's design and specifications as written by the purchaser.

The main purpose of a Fire Department pumper is to provide adequate pressure for fire streams. The water it pumps may come from its water tank, a fire hydrant, or an impounded supply.

ACCORDING TO THE PREVIOUS STATEMENT

1. Of the following, which is NOT a recognized capacity of pump for Fire Department pumpers:
   A. 1000 GPM.
   B. 1250 GPM.
   C. 1500 GPM.
   D. 1750 GPM.

   ANSWER = D

2. Fire Department pumpers must deliver 100% of their rated capacity at a pressure of:
   A. 100 PSI.
   B. 150 PSI.
   C. 200 PSI.
   D. 250 PSI.

   ANSWER = B

3. Fire Department pumpers must deliver 70% of their rated capacity at a pressure of:
   A. 100 PSI.
   B. 150 PSI.
   C. 200 PSI.
   D. 250 PSI.

   ANSWER = C

4. Fire Department pumpers must deliver 50% of their rated capacity at a pressure of:
A. 100 PSI.
B. 150 PSI.
C. 200 PSI.
D. 250 PSI.

ANSWER = D

5. Fire pumpers must be provided with adequate:
A. Inlet and discharge pump connections.
B. Pump and engine controls.
C. Gauges and other instruments.
D. All of the above.

ANSWER = D

6. A Fire Department triple combination pumper must have all of the following, EXCEPT:
A. Fire pump.
B. Water tank.
C. Hose compartment.
D. Hose.

ANSWER = D

7. The main purpose of a Fire pumper is to:
A. Deliver Firefighters to fire incidents.
B. Provide adequate pressure for fire streams.
C. Transport tools and equipment.
D. Transport water.

ANSWER = B

STATEMENT

The process of raising a ladder where it is needed will not in itself extinguish fire, but a well-positioned ladder becomes a means by which other operations can be performed. Fire Service ladders are essential for a rescue procedure, and carrying firefighting tools and appliances need to be carried above ground level. Teamwork, smoothness, and rhythm are necessary when raising and lowering Fire Department ladders, if speed and accuracy are to be developed. A knowledge of how ladders are mounted on fire apparatus, how they can be removed and used, along with an understanding of safety, spacing, and climbing techniques, is essential. These preliminary methods and skills provide a background for handling ladders. The recognized specifications for ladders recommend that they be made of metal.

ACCORDING TO THE PREVIOUS STATEMENT

1. Raising a ladder where it is needed in itself:
   A. Will extinguish a fire.
   B. Allow other operations to be performed.
   C. Both A and B.
   D. none of the above.

   ANSWER = B

2. Ladders are essential for all of the following, EXCEPT:
   A. Rescue procedures.
   B. Delivering tools and equipment above ground level.
   C. In many Fire Department operations.
   D. In all Fire Department operations.

   ANSWER = D

3. When raising and lowering ladders, it is necessary to have:
   A. Teamwork.
   B. Rhythm.
   C. Smoothness
   D. All of the above.

   ANSWER = D

4. Preliminary methods and skills that provide the knowledge and background for handling ladders include all of the following, EXCEPT:
   A. How ladders are mounted on apparatus.
   B. How ladders may be removed from apparatus.
   C. Where ladders are purchased.
   D. Understanding safety, spacing, and climbing techniques.

   ANSWER = C

5. Recognized specifications for ladders recommend that Fire Department ladders be made of:
   A. Birch.
   B. Oak.
   C. Ash.
   D. Metal.

   ANSWER = D

STATEMENT

Flash point is the lowest temperature of a liquid at which it gives off vapor which forms an ignitible mixture with air at the surface of the liquid, or within the vessel it is in. This is the lowest temperature that a flame will occur with a heat source. Fire will not continue without a heat source. Flash point is the ignitible temperature of a liquids vapor. The flash point more than any other physical property determines the hazard of a liquid. The flash point is at a temperature of less than 5 degrees F. below fire point. Fire point is the temperature that a liquid is able to continue to burn without an outside heat source. Fire point is the lowest temperature of at liquid at which vapors are evolved fast enough to support continued combustion.

ACCORDING TO THE PREVIOUS STATEMENT

1. Flash point of a liquid is:
   A. The lowest temperature that a flame will occur, with a heat source.
   B. The lowest temperature that a flame will occur, without a heat source.
   C. The lowest temperature that a flame will continue without an outside heat source.
   D. None of the above.

   ANSWER = A

2. Fire point is of a liquid is:
   A. The lowest temperature that a flame will occur, with a heat source.
   B. The lowest temperature that a flame will occur, without a heat source.
   C. The lowest temperature that a flame will continue, without an outside heat source.
   D. None of the above.

   ANSWER = C

3. Flash point of a liquid is:
   A. More than 5 degrees F. above fire point.
   B. Less than 5 degrees F. below fire point.
   C. The same temperature as fire point.
   D. Plus or minus 5 degrees of fire point.

   ANSWER = B

4. Flash point of a liquid is:
   A. The un-ignitable temperature of a liquids vapor.
   B. The continuously burning temperature of a liquids vapor, without a heat source.
   C. The ignitable temperature of a liquids vapor.
   D. Both B and C.

   ANSWER = C

5. The flash point of a liquid, more than any other physical property of a liquid determines:
   A. The temperature of the liquid.
   B. The density of the liquid.
   C. The heat generation of the liquid.
   D. The hazard of the liquid.
   ANSWER = D

STATEMENT

From one City to another, from one period of time to another, there are many factors that will remain constant in the evaluation of the efficiency of a Fire Department. There is a restricted usefulness in the unadjusted loss per a dollar amount valuation. It may be concluded that high Fire Department procedural expenditures could lean towards the association with a low fire loss. Statistical analysis of the loss and cost data in many Cities did not reveal any such relationship. The deficiency of such a correlation, even though to some degree due to failure to make the most constructive disbursement of fire protection resources, must be assigned at least in part to the obscuring result of differences in the natural, physical, and moral factors which affect fire risk.

ACCORDING TO THE PREVIOUS STATEMENT

1. Cities which spend large amounts of their resources on their Fire Departments:
   A. Will have higher fire losses than those that commit lesser sums to their Fire Department.
   B. Will not have the same total property valuation as those that commit lesser sums on their Fire Departments.
   C. Usually will not have lower fire losses than those that spend lesser sums on their Fire Departments.
   D. Usually will have lower fire losses than those that spend lesser sums on their Fire Departments.

   ANSWER = C

2. It can be determined that the unadjusted loss per a dollar amount is useful in comparing the Fire Departments of various Cities.
   A. Only if appropriately supervised experimental circumstances can be secured.
   B. Only if variations are made for other factors which affect fire loss.
   C. Only if the cities are of equal area and population.
   D. None of the above.

   ANSWER = B

3. A basis for failure to secure the anticipated correlation between Fire Department expenses and fire loss data, is:
A. The inefficiency of some Fire Department operations.
B. The constant changing dollar appraisal of property.
C. The statistical mistakes made by Fire Investigators.
D. The disturbing effect of swift technological modernization.

ANSWER = A

4. Of the following factors, the one that will most adequately reflect the fire risk in the unadjusted loss per dollar valuation, is:
A. The most common type of buildings within the cities boundaries.
B. The physical peculiarities of the City.
C. The Fire Department expenses of the city.
D. The total value of properties within the Cities boundaries.

ANSWER = D

STATEMENT

B.T.U. stands for British Thermal Unit. B.T.U. is the amount of heat required to raise one pound of water 1 degree F., at atmospheric pressure. To convert one pound of ice at 32 degrees F. to steam at 212 degrees F. requires 1293.7 B.T.U.'s, ice to water requires 143.4 B.T.U.'s, water to steam requires 970.3 B.T.U.'s, to raise 32 degrees F. to 212 degrees F. requires 180 B.T.U.'s. One gallon of water will absorb about 8000 B.T.U.'s. The number of B.T.U.'s required to raise one pound of a substance 1 degree F. is referred to as the specific heat. Specific heat is referred to as the absorption of heat.

ACCORDING TO THE PREVIOUS STATEMENT

1. B.T.U. stands for:
A. Best temperature used.
B. British Training Underwriters.
C. British thermal unit.
D. Better than usual.

ANSWER = C

2. At atmospheric pressure the B.T.U. is equal to the amount of heat required to:
A. Lower one pound of water 1 degree F.
B. Raise one pound of water 1 degree F.
C. Raise one gallon of water 1 degree F.
D. Lower one gallon of water 1 degree F.

ANSWER = B

3. 1293.7 B.T.U.'s are required to:
   A. Convert one pound of ice at 32 degrees F. to steam at 212 degrees F.
   B. Convert ice to water.
   C. Convert water to steam.
   D. Raise 32 degrees F. to 212 degrees F.

   ANSWER = A

4. 143.4 B.T.U.'s are required to:
   A. Convert one pound of ice at 32 degrees F. to steam at 212 degrees F.
   B. Convert ice to water.
   C. Convert water to steam.
   D. Raise 32 degrees F. to 212 degrees .

   ANSWER = B

5. 970.3 B.T.U.'s are required to:
   A. Convert one pound of ice at 32 degrees F. to steam at 212 degrees F.
   B. Raise 1 pound of substance 1 degree F.
   C. Raise 32 degrees F. to 212 degrees F.
   D. Convert water to steam.

   ANSWER = D

6. One gallon of water will absorb how many B.T.U.'s ?
   A. About 8 B.T.U.'s.
   B. About 800 B.T.U.'s.
   C. About 8000 B.T.U.'s.
   D. An undetermined amount of B.T.U.'s.

   ANSWER = C

STATEMENT

The fire triangle is a three sided figure that represents three factors necessary for combustion. The three factors necessary for combustion are: oxygen, heat, and fuel. The fire tetrahedron represents the four elements that are required by fire. The four elements required by a fire are: fuel, heat, oxygen, and uninhibited chain reaction.

ACCORDING TO THE PREVIOUS STATEMENT

1. The fire triangle is a three sided figure that represents three factors necessary for:
   A. Fire extinguishment.
   B. Fire safety.
   C. Fire combustion.
   D. Fire prevention.

   ANSWER = C

2. The fire triangle is a three sided figure that represents the three factors that are necessary for combustion, which of the following is not one of these factors.
A. Oxygen.
B. Carbon monoxide.
C. Heat.
D. Fuel.

ANSWER = B

3. What does the fire tetrahedron represent?
A. The three factors required for combustion.
B. The three factors required for fire extinguishment.
C. The four factors required for fire extinguishment.
D. The four elements required by fire.

ANSWER = D

4. All but one of the following are elements required by a fire, which one is it?
A. Fuel.
B. Heat.
C. Inhibited chain reaction.
D. Uninhibited chain reaction.
E. Oxygen.

ANSWER = C

5. One of the following is not a factor of the fire triangle, which one is it?
A. Fuel.
B. Heat.
C. Oxygen.
D. Inhibited chain reaction.

ANSWER = D

STATEMENT

Carbon monoxide is the most hazardous component in most fire gases. It combines with hemoglobin in the blood 210 times more readily than oxygen and it robs the blood of oxygen. 1.3% will cause unconsciousness in 2 or 3 breaths and death in a few minutes. Concentrations of more than .05% of carbon monoxide are dangerous. Most fire related deaths occur from carbon monoxide. Carbon monoxide is lighter than air.

Carbon monoxide is produced by each fire but is produced in larger quantities in fires of a smoldering nature because carbon monoxide is a product of incomplete combustion. Carbon monoxide is considered one of smokes deadly trio, along with hydrogen sulfide and hydrogen cyanide.

ACCORDING TO THE PREVIOUS STATEMENT

1. The chief danger in most fire gases is:
A. Carbon dioxide.
B. Carbon monoxide.
C. Hydrogen sulfide.
D. Hydrogen cyanide.

ANSWER = B

2. Most fire related deaths occur from:
A. Carbon Monoxide.
B. Carbon Dioxide.
C. Hydrogen sulfide.
D. Hydrogen cyanide.

ANSWER = A

3. Carbon monoxide is produced by each fire, but is produced in larger quantities in:
A. Fires of a free burning nature.
B. Fires in an outside area.
C. Fires of a smoldering nature.
D. A and B only.

ANSWER = C

4. Carbon monoxide is:
A. Heavier than air.
B. Heavier than water.
C. Lighter than air.
D. None of the above.

ANSWER = C

5. Carbon monoxide will do all of the following, EXCEPT?
A. Be absorbed by the blood stream 210 times more readily than oxygen.
B. Rob the blood of oxygen.
C. Cause death.
D. Cause oxygen to be absorbed in the blood stream at a faster rate than itself.

ANSWER = D

STATEMENT

When a Firefighter is fighting a fire with the use of water and the heating effect of the fire tends to be greater than the cooling effect of the water, the water will dissipate and evaporate as steam.

ACCORDING TO THE PREVIOUS STATEMENT

1. When fighting fire with water:
   A. Fighting a fire with water is ineffective.
   B. Fire may dissipate water.
   C. Water and steam are the same objects.
   D. The heating effect of fire varies conversely with the amount of water that is used.

   ANSWER = B

2. When fighting fire with water:
   A. The heating effect of the fire may be greater than the cooling effect of the water.
   B. The cooling effect of the water is always greater than the heating effect of fire.
   C. The heating effect of fire is always greater than the cooling effect of water.
   D. None of the above.

   ANSWER = A

3. When fighting fire with water:
   A. Water will never turn to steam.
   B. Water will always turn to steam.
   C. Water will sometimes turn to steam.
   D. none of the above.

   ANSWER = C

STATEMENT

Report to Captain Bruce Sutton in Suite #222 at 0800 hours on Saturday morning. He will give you your first assignment. Sometime during your first two weeks as a Firefighter, report to Mrs. Smith in the Personnel Office. She will want you to complete some paper work for the employment records. Your working hours for the first eight weeks will be from 0800 hours to 1800 hours for four days a week. You will get time and a half for all overtime that is required. For the first year as a Firefighter, you will be on a probationary status during which time you will not have any vacation time. At completion of the probationary period you will be granted 7 shifts per year off for vacation, and you will be granted 6 shifts per year off for sick leave.

ACCORDING TO THE PREVIOUS STATEMENT

1. What time and day should you report to Captain Sutton?
   A. 0700 hours on Friday.
   B. 0700 hours on Saturday.
   C. 0800 hours on Friday.
   D. 0800 hours on Saturday.

   ANSWER = D

2. When are you supposed to go to the Personnel Office?
   A. 0800 hours on Saturday.
   B. Sometime within your first week as a as a Firefighter.
   C. Sometime within your first two weeks as a Firefighter.
   D. Sometime within your first year probationary period as a Firefighter.

   ANSWER = C

3. What is the name of the person that you are to report to at the Personnel Office?
   A. Mrs. Jones.
   B. Miss. Jones.
   C. Mrs. Smith.
   D. Miss Smith.

   ANSWER = C

4. What are your working hours for the first eight weeks as a Firefighter?
   A. 0800 hours to 1800 hours.
   B. 0800 hours to 1600 hours.
   C. 0700 hours to 1800 hours.
   D. 0700 hours to 1600 hours.

   ANSWER = A

5. After your first year as a Firefighter, how many shifts will you be granted for vacation?
   A. 5 shifts.
   B. 6 shifts.
   C. 7 shifts.
   D. 8 shifts.

   ANSWER = C

6. After your first year as a Firefighter, how much time will you be granted for sick leave?
   A. 5 shifts.
   B. 6 shifts.
   C. 7 shifts.
   D. 8 shifts.

   ANSWER = B

VOCABULARY/VERBAL ABILITY

Entrance Firefighter exams will have VOCABULARY/VERBAL ABILITY TEST to discover a candidates knowledge of words.

When encountering vocabulary test: select the most nearly correct word from among the provided selections. Many times a synonym provided may not be the specific word you would use. But if it is the best of the choices it probably is the correct answer.

Watch out for clues, such as prefixes and suffixes of words. Using the word in a familiar context will help you.

Work rapidly and skim the potential answers to determine promptly the correct choice. Take time out to investigate individual choices only if the words are unfamiliar or troublesome to you.

You will not find exact definitions for the key words. You must use your ability to think and reason in order to select the best answer.

You should know if you are to select a synonym or antonym.

Remember don't panic when you see an unfamiliar word. Most of the words are regularly used in everyday use. Remain calm and you will be able to recognize words and their meanings.

1. **BARGAIN** means most nearly:
   A. agreement.
   B. debt.
   C. routine.
   D. design.

   ANSWER = A

2. **ADJOURN** means most nearly:
   A. start.
   B. handle.
   C. attend.
   D. complete.

   ANSWER = D

3. **CONSOLE** means most nearly:
   A. convey.
   B. find.
   C. reassure.
   D. scold.

   ANSWER = C

4. **MERIT** means most nearly:
A. deficiency.
B. warrant.
C. hope.
D. require.

ANSWER = B

5. **CORROBORATION** means most nearly:
A. accumulate.
B. reduction.
C. confirmation.
D. expenditure.

ANSWER = C

6. **OPTION** means most nearly:
A. alternative.
B. use.
C. blame.
D. error.

ANSWER = A

7. **ZEAL** means most nearly:
A. kindness.
B. faith.
C. integrity.
D. enthusiastic.

ANSWER = D

8. **SLOTH** means most nearly:
A. hatred.
B. lethargic.
C. distress.
D. selfishness.

ANSWER = B

9. **DEFAMATION** means most nearly:
A. debt.
B. embezzlement.
C. slander.
D. contamination.

ANSWER = C

10. **PERTINENT** means most nearly:
A. relevant.
B. persuade.
C. foolproof.
D. careful.

ANSWER = A

11. **IMPROMPTU** means most nearly:
    A. laughable.
    B. attractive.
    C. insulting.
    D. spontaneous.

    ANSWER = D

12. **ENDOW** means most nearly:
    A. death.
    B. abandon.
    C. hand down.
    D. bless.

    ANSWER = C

13. **ALTERATION** means most nearly:
    A. prayer.
    B. change.
    C. cue.
    D. temperance.

    ANSWER = B

14. **ABSORB** means most nearly:
    A. consume.
    B. spread.
    C. purge.
    D. repair.

    ANSWER = A

15. **PERCEPTIVE** means most nearly:
    A. astute.
    B. vicious.
    C. dense.
    D. fair.

    ANSWER = A

16. **CANDID** means most nearly:
    A. correct.
    B. direct.
    C. hasty.
    D. careful.

    ANSWER = B

17. **BANQUET** means most nearly:
    A. benefaction.
    B. hall.
    C. feast.
    D. surprise.

    ANSWER = C

18. **INFILTRATE** means most nearly:
    A. deflate.
    B. rebound.
    C. indentation.
    D. penetrate.

    ANSWER = D

19. **INTERFERE** means most nearly:
    A. stagger.
    B. joking.
    C. meddling.
    D. inquisitive.

    ANSWER = C

20. **THRUST** means most nearly:
    A. conclude.
    B. pursue.
    C. vigil.
    D. launch.

    ANSWER = D

21. **FRAIL** means most nearly:
    A. fragile.
    B. tired.
    C. pretty.
    D. vain.

    ANSWER = A

22. **HABITUAL** means most nearly:
    A. rare.
    B. drastic.
    C. never.
    D. routine.

    ANSWER = D

23. **PROMPT** means most nearly:
    A. courteous.
    B. punctual.
    C. considerate.
    D. immaculate.

    ANSWER = B

24. **ADEQUATE** means most nearly:
    A. ample.
    B. less.
    C. too much.
    D. too little.

    ANSWER = A

25. **JUSTIFY** means most nearly:
   A. shield.
   B. appreciate.
   C. complete.
   D. defend.

   ANSWER = D

26. **INTRUDE** means most nearly:
   A. overtake.
   b. duplicate.
   C. trespass.
   D. persist.

   ANSWER = C

27. **INVARIABLY** means most nearly:
   A. sometimes.
   B. periodically.
   C. consistently.
   D. never.

   ANSWER = C

28. **RESOURCES** means most nearly:
   A. portion.
   B. assets.
   C. compromise.
   D. maraud.

   ANSWER = B

29. **WARRANT** means most nearly:
   A. verify.
   B. investigate.
   C. realize.
   D. limit.

   ANSWER = A

30. **OFFENSIVE** means most nearly:
   A. vision.
   B. irate.
   C. unsuitable.
   D. dauntless.

   ANSWER = C

31. **SLENDER** means most nearly:
   A. overgrown.
   B. seductive.
   C. stormy.
   D. thin.

   ANSWER = D

32. **TRANQUILIZING** means most nearly:
   A. calming.
   B. supernatural.
   C. solemn.
   D. harmonize.

   ANSWER = A

33. **OBJECTIVE** means most nearly:
   A. competent.
   B. unbiased.
   C. faithful.
   D. compassionate.

   ANSWER = B

34. **CRAFTY** means most nearly:
   A. humorous.
   B. intriguing.
   C. clever.
   D. wicked.

   ANSWER = C

35. **CONTEMPLATE** means most nearly:
   A. envision.
   B. appreciate.
   C. invalidate.
   D. disguise.

   ANSWER = A

36. **FALLACY** means most nearly:
   A. clemency.
   B. evidence.
   C. liability.
   D. erroneous.

   ANSWER = D

37. **INTENTION** means most nearly:
   A. desire.
   B. objective.
   C. speculation.
   D. apprehension.

   ANSWER = B

38. **ACUTE** means most nearly:
   A. painful.
   B. sharp.
   C. authentic.
   D. serious.

   ANSWER = D

39. **DEBONAIR** means most nearly:
   A. charming.
   B. dissolute.
   C. dainty.
   D. exorbitant.

   ANSWER = A

40. **HABITAT** means most nearly:
   A. system.
   B. posture.
   C. dwelling.
   D. wardrobe.

   ANSWER = C

41. **GRATIFYING** means most nearly:
   A. offensive.
   B. considerate.
   C. revelation.
   D. pleasing.

   ANSWER = D

42. **CANDID** means most nearly:
   A. swift.
   B. direct.
   C. contention.
   D. kind.

   ANSWER = B

43. **ABBREVIATED** means most nearly:
   A. abridged.
   B. actual.
   C. saturated.
   D. established.

   ANSWER = A

44. **CAMOUFLAGE** means most nearly:
   A. retrieve.
   B. plagiarize.
   C. conceal.
   D. intercept.

   ANSWER = C

45. **INCONSEQUENTIAL** means most nearly:
   A. unforeseen.
   B. trivial.
   C. unnecessary.
   D. unequivocal.

   ANSWER = B

46. **PROSCRIBE** means most nearly:
   A. promote.
   B. administer.
   C. produce.
   D. forbid.

   ANSWER = D

47. **BOUT** means most nearly:
   A. boot.
   B. match.
   C. disease.
   D. craft.

   ANSWER = B

48. **CONSOLIDATE** means most nearly:
   A. combine.
   B. reinforce.
   C. identify.
   D. classify.

   ANSWER = A

49. **LOQUACIOUS** means most nearly:
   A. leafy.
   B. redolence.
   C. talkative.
   D. lightheaded.

   ANSWER = C

50. **EXTOL** means most nearly:
   A. praise.
   B. release.
   C. split.
   D. rush.

   ANSWER = A

51. **CONFRONT** means most nearly:
   A. challenge.
   B. determine.
   C. unlock.
   D. relinquish.

   ANSWER = A

52. **FORTIFY** means most nearly:
   A. excavate.
   B. extend.
   C. support.
   D. reduce.

   ANSWER = C

53. **EXPAND** means most nearly:
   A. remodel.
   B. increase.
   C. construct.
   D. absorb.

   ANSWER = B

54. **ACCENTUATE** means most nearly:
   A. acquaint.
   B. feature.
   C. omit.
   D. protest.

   ANSWER = B

55. **RELUCTANT** means most nearly:
   A. consistent.
   B. extreme.
   C. disinclined.
   D. repose.

   ANSWER = C

56. **LOOMING** means most nearly:
   A. threatening.
   B. distressed.
   C. transient.
   D. significant.

   ANSWER = A

57. **BRAVERY** means most nearly:
   A. fervor.
   B. boldness.
   C. independence.
   D. potency.

   ANSWER = B

58. **CLIENTELE** means most nearly:
   A. attendants.
   B. occupants.
   C. congregation.
   D. patrons.

   ANSWER = D

59. **RELINQUISH** means most nearly:
   A. require.
   B. surrender.
   C. reserve.
   D. respite.

   ANSWER = B

60. **ANTECEDE** means most nearly:
   A. introduce.
   B. reluctance.
   C. cavalcade.
   D. ceremony.

   ANSWER = A

61. **AMIABLE** means most nearly:
   A. offensive.
   B. likeable.
   C. emotional.
   D. confidential.

   ANSWER = B

62. **CORRUPT** means most nearly:
   A. majestic.
   B. vigil.
   C. detached.
   D. immoral.

   ANSWER = D

63. **UNCANNY** means most nearly:
   A. remarkable.
   B. juvenile.
   C. dishonest.
   D. spontaneous.

   ANSWER = A

64. **COGNIZANT** means most nearly:
   A. concerned.
   B. unsuitable.
   C. conscious.
   D. thrilled.

   ANSWER = C

65. **PROVOKE** means most nearly:
   A. conceal.
   B. thwart.
   C. vicious.
   D. incite.

   ANSWER = D

66. **ERADICATE** means most nearly:
   A. demonstrate.
   B. remove.
   C. detain.
   D. disguise.

   ANSWER = B

67. **INJURY** means most nearly:
   A. Damage.
   B. threat.
   C. decease.
   D. sorrow.

   ANSWER = A

68. **PROLOGUE** means most nearly:
   A. table of contents.
   B. presentation.
   C. opening.
   D. encasement.

   ANSWER = C

69. **PRECARIOUS** means most nearly:
   A. illusive.
   B. sluggish.
   C. exciting.
   D. risky.

   ANSWER = D

70. **CONTROVERSY** means most nearly:
   A. conflict.
   B. dialogue.
   C. presentation.
   D. entertainment.

   ANSWER = A

71. **ABRUPTLY** means most nearly:
   A. soon.
   B. obscure.
   C. hasty.
   D. noisily.

   ANSWER = C

Another type of vocabulary test will give a word in a sentence and then ask the candidate to select another word from a list that means most nearly the same as the word used in the sentence.

EXAMPLES:
IN THE FOLLOWING SENTENCES THE WORD THAT IS IN **BOLD/CAPITAL** LETTERS MEANS MOST NEARLY?
(In these situations only a possible correct choice is given)

1. The breakdown of the machine was due to a defective **GASKET**.

   ANSWER = SEALER

2. The garden contains a **PROFUSION** of flowers.

   ANSWER = ABUNDANCE

3. The person was **CAJOLED** into signing the contract.

   ANSWER = COAXED

4. Noise from the **PNEUMATIC** hammer bothered the man.

   ANSWER = AIR

5. It is impossible to **MISCONSTRUE** my letter.

   ANSWER = MISINTERPRET

6. The mother **ADMONISHED** the child for his behavior.

   ANSWER = WARNED

7. The foreman would not approve the job since it was out of **PLUMB**.

   ANSWER = NOT VERTICAL

8. The speaker's decree was met with general **DERISION**.

   ANSWER = RIDICULE

9. The speaker's decree was **IRRELEVANT**.

   ANSWER = UNCONNECTED

10. The baseball bat was considered a **LETHAL** weapon.

    ANSWER = DEADLY

11. John was appointed **PROVISIONAL** Captain.

    ANSWER = TEMPORARY

12. The Fire Chief used his **PREROGATIVES** in moderation.

    ANSWER = PRIVILEGES

13. The Firefighter used his **INITIATIVE** to complete the task.

    ANSWER = MOTIVATION

14. The Fire Engineer used his **SEASONING** in order to get through the tight situation.

    ANSWER = TRAINING

And still another type of vocabulary exam will give the candidate a relationship between two words and then ask the candidate to select from a list a relationship from a third given word.

IN THE FOLLOWING SENTENCES, THE WORDS IN **BOLD/CAPITAL** LETTERS ARE RELATED MOST NEARLY TO A WORD IN THE LIST OF GIVEN CHOICES!

EXAMPLES:

1. APRIL is to MONTH as **MONDAY** is to:
   A. minute
   B. hour.
   C. day.
   D. week.

   ANSWER = C

2. ORE is to METAL as **HIDE** is to:
   A. leather.
   B. belt.
   C. plastic.
   D. shoe.

   ANSWER = A

3. GRAIN is to OAT as **VEGETABLE** is to:
   A. wheat.
   B. carrot.
   C. rose.
   D. robin.

   ANSWER = B

4. HUNGER is to NOURISHMENT as **FATIGUE** is to:
   A. work.
   B. play.
   C. sickness.
   D. rest.

   ANSWER = D

5. SCALE is to POUNDS as **RULER** is to:
   A. line.
   B. inches.
   C. length.
   D. measurement.

   ANSWER = B

6. PHYSICIAN is to PATIENT as **LAWYER** is to:
   A. client.
   B. legal representation.
   C. license.
   D. court.

   ANSWER = A

7. FELONY is to CRIME as **EAGLE** is to :
   A. hawk.
   B. fish.
   C. bird
   D. feather.

   ANSWER = C

Vocabulary exams will also ask the candidate to select the opposite meaning of a word:

EXAMPLES:

1. ALOOF means the opposite of:
   A. mute.
   B. loquacious.
   C. silent.
   D. reserved.

   ANSWER = B

2. PHILANTHROPIC means the opposite of:
   A. grand.
   B. microscopic.
   C. stingy.
   D. powerful.

   ANSWER = C

3. SINISTER means the opposite of:
   A. availing.
   B. unsightly.
   C. factual.
   D. undemanding.

   ANSWER = A

4. AGGRESSIVE means the opposite of:
   A. hostile.
   B. contentious.
   C. tranquil.
   D. forceful.

   ANSWER = C

5. FLAMBOYANT means the opposite of:
   A. crucial.
   B. flashy.
   C. somber.
   D. loud.

   ANSWER = C

6. JADED means the opposite of:
   A. corrupt.
   B. rejuvenated.
   C. shoddy.
   D. nasty.

   ANSWER = B

7. GROVELING means the opposite of:
   A. cringe.
   B. servile.
   C. admirable.
   D. decline.

   ANSWER = C

8. LOATHE means the opposite of:
   A. despise.
   B. cherish.
   C. disdain.
   D. execrate.

   ANSWER = B

SECTION 4
SPELLING / GRAMMAR

SPELLING QUESTIONS

Many ENTRANCE LEVEL FIREFIGHTER EXAMS will have a portion of the exam devoted to spelling.

When encountering the spelling portion of an exam: read all the word choices carefully before deciding on your answer. Scan reading is not recommended on spelling test.

Watch out for words that are exceptions to the rules of spelling. Apply as many of the spelling rules that you can and don't let the exception words scare you!

Have a positive attitude and work quickly and methodically. Concentrate on how a word is spelled and not on its definition.

The answers will not follow any special pattern. Each question and answer will be independent of the other questions.

The following examples are characteristic of the degree of spelling that is required on Firefighter exams.

CHOOSE THE CORRECT SPELLING OF THE FOLLOWING WORDS:

1.
 A. abdoeman.
 B. abdomen.
 C. abdoman.
 D. none of the above.

 ANSWER = B

2.
 A. acelerator.
 B. accelorator.
 C. accelerator.
 D. none of the above.

 ANSWER = C

3.
 A. accordion.
 B. acordyian.
 C. accordyon.
 D. none of the above.

 ANSWER = A

4.
 A. airial.
 B. aerial.
 C. aereal.
 D. none of the above.

 ANSWER = B

5.
 A. altornator.
 B. altarnator.
 C. altanator.
 D. none of the above.

 ANSWER = D

6.
 A. baffle.
 B. bafol.
 C. baffol.
 D. none of the above.

 ANSWER = A

7.
 A. barameter
 B. barometor.
 C. barometer.
 D. none of the above.

 ANSWER = C

8.
 A. battalion.
 B. batallion.
 C. batalloun.
 D. none of the above.

 ANSWER = A

9.
 A. bimetalic.
 B. bimetallic.
 C. bymetallic.
 D. none of the above

 ANSWER = B

10.
 A. bowlyn.
 B. bowline.
 C. bowlynn.
 D. none of the above.

 ANSWER = B

11.
   A. campain.
   B. campaien.
   C. campaign.
   D. none of the above.

   ANSWER = C

12.
   A. cantilever.
   B. cantolever.
   C. cantalever.
   D. none of the above.

   ANSWER = A

13.
   A. captian.
   B. captain.
   C. captin.
   D. none of the above.

   ANSWER = B

14.
   A. carpenter.
   B. carponder.
   C. carponter.
   D. none of the above.

   ANSWER = A

15.
   A. cardopulmonary.
   B. cardyopulmonary.
   C. cardiopulmonary.
   D. none of the above.

   ANSWER = C

16.
   A. dalmation.
   B. dalmatian.
   C. dalmatain.
   D. none of the above.

   ANSWER = B

17.
   A. deluge.
   B. delouge.
   C. daluge.
   D. none of the above.

   ANSWER = A

18.
   A. difusson.
   B. diffusson.
   C. diffusion.
   D. none of the above.

   ANSWER = C

19.
   A. directory.
   B. directery.
   C. directary.
   D. none of the above.

   ANSWER = A

20.
   A. durable.
   B. dureble.
   C. durible.
   D. none of the above.

   ANSWER = A

21.
   A. elevation.
   B. elavation.
   C. elivation.
   D. none of the above.

   ANSWER = A

22.
   A. emergoncy.
   B. emergancy.
   C. emergincy.
   D. none of the above.

   ANSWER = D

23.
   A. endurence.
   B. endurince.
   C. endurance.
   D. none of the above.

   ANSWER = C

24.
   A. expander.
   B. expandor.
   C. expandar.
   D. none of the above.

   ANSWER = A

25.
   A. extinguesh.
   B. extinguish.
   C. extingwish.
   D. none of the above.

   ANSWER = B

26.
   A. flamability.
   B. flammability.
   C. flammibility.
   D. none of the above.

   ANSWER = B

27.
   A. flodlight.
   B. floodlight.
   C. floodlite.
   D. none of the above.

   ANSWER = B

28.
   A. fundamental.
   B. fundemental.
   C. fundamentol.
   D. none of the above.

   ANSWER = A

29.
   A. fritcion.
   B. frction.
   C. friction.
   D. none of the above.

   ANSWER = C

30.
   A. fuesible.
   B. fusable.
   C. fusible.
   D. none of the above.

   ANSWER = C

31.
   A. gasoline.
   B. gasoleen.
   C. gasolien.
   D. none of the above.

   ANSWER = A

32.
 A. genorator.
 B. generator.
 C. generrator.
 D. none of the above.

 ANSWER = B

33.
 A. govonor.
 B. governor.
 C. governar.
 D. none of the above.

 ANSWER = B

34.
 A. gravaty.
 B. gravety.
 C. gravity.
 D. none of the above.

 ANSWER = C

35.
 A. gutter.
 B. guttor.
 C. guttar.

 ANSWER = A

36.
 A. hazzard.
 B. hazzord.
 C. hazard.
 D. none of the above.

 ANSWER = C

37.
 A. horizontal.
 B. horazontal.
 C. horizontol.
 D. none of the above.

 ANSWER = A

38.
 A. hydrent.
 B. hydrant.
 C. hydrunt.
 D. none of the above.

 ANSWER = B

39.
   A. hydrolic.
   B. hydralic.
   C. hydraullic.
   D. none of the above.

   ANSWER = D

40.
   A. hydremeter.
   B. hydrameter.
   C. hydrometer.
   D. none of the above.

   ANSWER = C

41.
   A. ignightable.
   B. igniteable.
   C. ignitable.
   D. none of the above.

   ANSWER = C

42.
   A. ignition.
   B. ignetion.
   C. ignitian.
   D. none of the above.

   ANSWER = A

43.
   A. imiscible.
   B. immiscible.
   C. immiscable.
   D. none of the above.

   ANSWER = B

44.
   A. immpeller.
   B. impellor.
   C. impellar.
   D. none of the above.

   ANSWER = D

45.
   A. impingement.
   B. impengement.
   C. impingment.
   D. none of the above.

   ANSWER = A

46.
   A. jaccnife.
   B. jackknife.
   C. jacknefe.
   D. none of the above.

   ANSWER = B

47.
   A. janator.
   B. janitoor.
   C. janitor.
   D. none of the above.

   ANSWER = C

48.
   A. jeopardize.
   B. jepordize.
   C. jepardize.
   D. none of the above.

   ANSWER = A

49.
   A. jostel.
   B. jostal.
   C. jostle.
   D. none of the above

   ANSWER = C

50.
   A. kerascene.
   B. keroscene.
   C. kerosene.
   D. none of the above.

   ANSWER = C

51.
   A. kindaling.
   B. kindoling.
   C. kindling.
   D. none of the above.

   ANSWER = C

52.
   A. kinitic.
   B. kenitic.
   C. kanitic.
   D. none of the above.

   ANSWER = D

53.
A. kilogram.
B. kilagram.
C. kilygram.
D. none of the above.

ANSWER = A

54.
A. knote.
B. knott.
C. knot.
D. none of the above.

ANSWER = C

55.
A. knowladge
B. knowledge.
C. knowlege.
D. none of the above.

ANSWER = B

56.
A. labrinth.
B. labyrnth.
C. laberinth.
D. none of the above.

ANSWER = D

57.
A. latont.
B. latint.
C. latent.
D. none of the above.

ANSWER = C

58.
A. lattitude.
B. latatude.
C. latitude.
D. none of the above.

ANSWER = C

59.
A. legitimate.
B. legitamate.
C. lagitimate.
D. none of the above.

ANSWER = A

60.
   A. leutenent.
   B. liutenant.
   C. leiutenant.
   D. none of the above.

   ANSWER = D

61.
   A. maxamum.
   B. maximum.
   C. maxumum.
   D. none of the above.

   ANSWER = B

62.
   A. mercontile.
   B. mercantile.
   C. mercuntile.
   D. none of the above.

   ANSWER = B

63.
   A. mezzine.
   B. mezanine.
   C. mezzanine.
   D. none of the above.

   ANSWER = C

64.
   A. millameter.
   B. millemmetor.
   C. millemetre.
   D. none of the above.

   ANSWER = D

65.
   A. multiple.
   B. multaple.
   C. multipel.
   D. none of the above.

   ANSWER = A

Another type of SPELLING TEST, will list a group of words, one of which is misspelled. The candidate will be required to select the word from the list that is misspelled.

EXAMPLES:

1. A. aisle   B. cemetary   C. courtesy   D. phlegm

    ANSWER = B, correct spelling = cemetery

2. A. hypocrisy   B. extraordinary   C. dogma   D. auxilliary

    ANSWER = D, correct spelling = auxiliary

3. A. intorsect   B. launch   C. approach   D. defective

    ANSWER = A, correct spelling = intersect

4. A. conferred   B. gigantac   C. synthetic   D. caution

    ANSWER = B, correct spelling = gigantic

5. A. narrow   B. dispute   C. bargan   D. rapidity

    ANSWER = C, correct spelling = bargain

6. A. adjourne   B. hazardous   C. meddling   D. inquisitive

    ANSWER = A, correct spelling = adjourn

7. A. inquisitive   B. liberate   C. habituall   D. spacious

    ANSWER = C, correct spelling = habitual

8. A. reliable   B. boulevard   C. portable   D. illitorate

    ANSWER = D, correct spelling = illiterate

9. A. permanent   B. abruptly   C. barracade   D. methodical

    ANSWER = C, correct spelling = barricade

10. A. punctual   B. penitrate   C. vegetation   D. juvenile

    ANSWER = B, correct spelling = penetrate

GRAMMAR

Many FIREFIGHTER ENTRANCE EXAMS will have a section which examines your knowledge and understanding of the basic rules of ENGLISH GRAMMAR.

When encountering the grammar portion of an examination: study the instructions and examples carefully. Even though every question will involve your consciousness of basic grammar, the questions will deviate in format.

Always study the sentences carefully. Recognize that the location of a comma or clause can alter the usage.

Watch for sentence fragments or incomplete sentences. They may look correct in a rapid reading, but remember that you are being tested for what is wholly correct.

Organize your time and do not spend to much time on any one question. If you can make an intelligent guess, make it or go on to the next question and come back later if you have time.

Remember that you are being requested to consider several things in each sentence, such as, spelling, and punctuation.

Remember to select the best answer. More than one answer may appear to be correct.

The following questions are examples of the components of GRAMMAR that candidates may be tested on, such as punctuation, articulation, and integral parts of speech.

In the following examples choose the sentence that is the most correct as far as grammar, usage, and punctuation:

1.
   A. Most of the firefighters' training was supervised by captains who's interest lay in training.
   B. Most of the firefighter's training was supervised by captains who's interest lay in training.
   C. Most of the firefighter's training was supervised by captains' whose interest lay in training.

ANSWER = C

2.
   A. As the reverberations of the siren increase, one of the neighborhood dogs starts to yowl.
   B. As the reverberations, of the siren increase, one of the neighborhood dogs start to yowl.
   C. As the reverberations of the siren increases, one of the neighborhood dogs start to yowl.

ANSWER = A

3.
   A. Bill yells at Donna that it is her, not he, who hollers.
   B. Bill yells at Donna that it is her, not he, who hollers.
   C. Bill yells at Donna that it is she, not him who hollers.

ANSWER = C

4.
   A. When a noble fireman trains, he feels real good.
   B. When a noble fireman trains, he feels really good.
   C. When a noble fireman trains, he really feels good.

ANSWER = C

5.
   A. The candidate asked the proctor what he should do with the examination booklet. Can you envision what he stated?
   B. The candidate asked the proctor what he should do with the examination booklet? can you envision what he stated?
   C. The candidate asked the proctor what he should do with the examination booklet. Can you envision what he stated.

ANSWER = A

6.
A. I remember the academy where I trained, while I lived alone, when I was younger.
B. I remember the academy, where I trained, while I lived alone when I was younger.
C. I remember the academy, where I trained, while I lived alone when I was younger.

ANSWER = B

7.
A. When the staff reports its verdict, some fireman will lose his poise.
B. When the staff reports their verdict, some firemen will lose there poise.
C. When the staff reports its verdict, some fireman will lose their poise.

ANSWER = A

8.
A. Many of the captains' drills were attended by firefighters who's interest lay in learning.
B. Many of the captain's drills were attended by firefighters who's interest lay in learning.
C. Many of the captains' drills were attended by firefighters whose interest lay in learning.
D. Many of the captain's drills were attended by firefighters whose interest lay in learning.

ANSWER = D

9.
A. I asked the instructor what I should do with this examination paper. Can you imagine what he said?
B. I asked the instructor what I should do with this examination paper? Can you imagine what he said.
C. I asked the instructor what I should do with this examination paper? Can you imagine what he said?
D. I asked the instructor what I should do with this examination paper? Can you imagine what he said!

ANSWER = A

10.
   A. Its in untried emergencies that a firefighter's composure receives its supreme test.
   B. It's in untried emergencies that a firefighter's composure receives its supreme test.
   C. It's in untried emergencies that a firefighters' composure receives its supreme test.
   D. It's in untried emergencies that a firefighter's composure receives its' supreme test.

   ANSWER = B

11.
   A. What you say may be different from me.
   B. What you say may be different from what I say.
   C. What you say may be different than me.
   D. What you say may be different from mine.

   ANSWER = B

12.
   A. It takes study to become a firefighter.
   B. It takes study before you can become a firefighter.
   C. It takes study in becoming a firefighter.
   D. It takes study about to become a firefighter.

   ANSWER = A

13.
   A. You people appreciate we firefighters as much as we firefighters appreciate you.
   B. You people appreciate we firefighters as much as us firefighters appreciate you.
   C. You people appreciate us firefighters as much as us firefighters appreciate you.
   D. You people appreciate us firefighters as much as we firefighters appreciate you.

   ANSWER = D

14.
   A. He use to visit when he was supposed to.
   B. He use to visit when he was suppose to.
   C. He used to visit when he was supposed to.
   D. He visits when he was supposed to.
   ANSWER = C

15.
   A. I saw the engineer and asked him for a hose, fitting, and nozzle.
   B. I saw the engineer, and asked him for a hose, fitting, and nozzle.
   C. I saw the engineer and asked him for a hose, fitting and nozzle.
   D. I saw the engineer asking for a hose, fitting, and nozzle.

   ANSWER = A

16.
   A. A short, old firefighter threw the heavy, soggy, salvage cover.
   B. A short, old firefighter, threw the heavy, soggy salvage cover.
   C. A short old firefighter threw the heavy, soggy salvage cover.
   D. A short old firefighter threw the heavy soggy salvage cover.

   ANSWER = C

17.
   A. We rarely eat meat at our firehouse. My captain being a vegetarian.
   B. We rarely eat meat at our firehouse my captain being a vegetarian.
   C. We rarely eat meat at our firehouse; my captain being a vegetarian.
   D. We rarely eat meat at our firehouse, my captain being a vegetarian.

   ANSWER = D

18.
   A. The commander has only one request. That you respond without delay.
   B. The commander has only one request: that you respond without delay.
   C. The commander has only one request that you respond without delay.
   D. The commander has only one request; that you respond without delay.

   ANSWER = B

19.
   A. The captain insisted that you and he were responsible for the mistakes of Mac and me.
   B. The captain insisted that you and him were responsible for the mistakes of Mac and me.
   C. The captain insisted that you and he were responsible for the mistakes of Mac and I.
   D. The captain insisted that you and him were responsible for the mistakes of Mac and I.

   ANSWER = A

20.
   A. Certainly, even the most experienced firefighter feels stress.
   B. Certainly even the most experienced firefighter feels stress.
   C. Certainly; even the most experience firefighter feels stress.
   D. Certainly: even the most experienced firefighter feels stress.

   ANSWER = A

The following examples will show a statement written incorrectly, followed by the correct form in bold type.

EXAMPLES:

1. Every country has their representative at the conference.

   Every country has its representative at the conference.

2. I read all but the last paragraphs of his essay.

   I read all but the last few paragraphs of his essay.

3. Both the classical and the general course prepares the student for college.

   Both the classical and the general course prepare the student for college.

4. After spending several weeks there, he wrote that he was enamored with the countryside.

   After spending several weeks there, he wrote that he was enamored of the countryside.

5. The crate of eggs, together with the dozen containers of milk, were punctured in scores of places.

   The crate of eggs, together with the dozen containers of milk, was punctured in scores of places.

6. I like those kind of scissors best.
   I like this kind of scissors best.

7. I feel badly because of the oversight.
   I feel bad because of the oversight.

8. He was an artist, a scholar, and he could talk!
   He was an artist, a scholar, and a talker.

9. Her brother graduated high school last year.
   Her brother graduated from high school last year.

10. In their own way, everybody has to carry on as best they can.
    In his own way, everybody has to carry on as best he can.

11. A special light will be required to inspect the engine.
    To inspect the engine, a special light will be required.

12. The shift before, my captain thinking of other matters thrust his hand into a fire.
    The shift before, my captain, thinking of other matters, thrust his hand into a fire.

13. Not wishing to hurt my captain's feelings, I told him that I was leaving, because I had a previous engagement.
    Not wishing to hurt my captain's feelings, I told him that I was leaving because I had a previous engagement.

14. Would captain Sutton have survived if he was less imaginative?
    Would captain Sutton have survived if he had been less imaginative?

15. The worst one of the problems which is confronting me concern morale.

  The worst one of the problems which are confronting me concerns morale.

16. When the oral board reports their decision, somebody will earn their promotion.

  When the oral board reports their decision, somebody will earn his promotion.

17. Off in the distance is the fire truck, but there's no signs of it yet.

  Off in the distance is the fire truck, but there are no signs of it yet.

18. Neither of the paramedics believe that the driver or passenger are alive.

  Neither of the paramedics believe that the driver or passenger is alive.

19. Expecting my crew to be on time, their tardiness seemed almost an insult.

  Expecting my crew to be on time, I regarded their tardiness almost as an insult.

20. When mixing it, the foam must be thoroughly blended.

  When being mixed, the foam must be thoroughly blended.

21. It is a thing of excitement, responsibility, and containing satisfaction.

  It is a thing of excitement, responsibility, and satisfaction.

22. If the captain was able, he would demand that the engine return to the station.

  If the captain were able, he would demand that the engine return to the station.

23. You first wash the salvage cover in water. Then hang it up to dry.

    First, wash the salvage cover in water. Then hang it up to dry.

24. The captain sometimes in a good mood gave the study material to others that he had composed.

    Sometimes in a good mood, the captain gave to others the study material that he had composed.

25. Even firefighters of uneasy disposition periodically feel an element of primitive enjoyment.

    Even firefighters of uneasy disposition periodically feel an element of primitive enjoyment.

26. I like those kind of scissors best.

    I like this kind of scissors best.

27. If he would have stated his case, the captain would have been more lenient.

    If he had stated his case, the captain would have been more lenient.

28. He was a firefighter, writer, and he could talk!

    He was a firefighter, writer, and a talker.

29. In their own way, everybody has to carry on as best they can.

    In his own way, everybody has to carry on as best he can.

30. History has seldom and perhaps will never again record such heroism.

    History has seldom recorded, and perhaps will never again record such heroism.

SECTION 5

MECHANICAL COMPREHENSION

EXPLANATION

The job of firefighting consist of many tasks that invoke mechanical performance, therefore, many ENTRANCE LEVEL FIRE DEPARTMENT EXAMINATIONS will have a section that will require the candidates to demonstrate their knowledge and abilities relating to MECHANICAL COMPREHENSION.

When encountering mechanical type questions: base your response on the information provided with each question, and nothing more! The correct answer for the question will only relate to the stated problem. Even if you consider that the information presented is not accurate or up to date, base your answer on it and it only.

Read each question carefully, paying particular attention to questions with drawings. You need to examine the questions, the diagrams, and the information that goes with some of the diagrams.

When solving problems involving illustrations, stay composed, even if the illustrations look unfamiliar difficult, or the questions appear obscure. Once you have studied the question in connection to the illustration, you will likely be able to figure out an answer.

THE FOLLOWING INFORMATION WILL ASSIST YOU IN SOLVING MECHANICAL PROBLEMS THAT YOU MIGHT ENCOUNTER DURING THE EXAM

GENERAL INFORMATION

BASIC THEORIES AND CONCEPTS

When two objects of the same size, but of different weights, are dropped from a given height, both objects will reach the ground at the same time because the effect of gravity is the same on both objects.

The air in the highest portion of a room is normally higher in temperature than the air near the floor because warmer air is lighter that cold air.

When using an auger bit to drill a hole through a piece of wood, it is a good idea to clamp a piece of scrap wood on the underside of the piece that is being worked on so as to prevent it from splintering.

An electric fuse is used to protect against overload.

Electrical conductors are generally made of copper.

An electron is a particle of negative electricity.

Any substance which offers a very high or very great resistance to the passage of electric current through it is called an insulator.

Water is a better conductor of electricity when its mineral content is high.

ELECTRICAL SERVICE ENTRANCE CONSIST OF:
1. Service drop, service head, and insulators.
2. Conduit or cable.
3. Weather proof meter and socket.
4. Entrance switch with circuit breakers.
5. Ground wire.

Electrical VOLTAGE is like PSI (pressure).

Electrical AMPERES is like GPM (rate of flow).

Electrical OHMS is like FL (friction loss, resistance).

ELECTRICAL AMPERES is the most hazardous component of a high voltage electrical circuit.

OHMS LAW:
1. Amperes = volts divided by ohms.
2. Volts   = amperes multiplied by ohms.
3. Ohms   = volts divided by amperes.

OVERLOADING an electrical circuit will cause overheating.

Maximum continuous current an individual may safely be subjected to is FIVE (5) MILLIAMPERES.

TRANSFORMERS reduce voltage.

There is an INCREASE of life hazard with the magnitude of voltage.

PROTONS are positive charged and NEUTRONS have no charge.

LOOSELY BOUND ELECTRONS are good insulators. (there are no perfect insulators).

In TRANSFORMER FIRES, let fire burn itself out.

IONIZED AIR is the most conductive of electrical charges.

In electrical distribution systems, PRIMARIES are the high voltage lines and SECONDARIES are the low voltage side of the system.

The voltage side of the transformer to home or place of business is equal to 600 VOLTS or less.

Firefighters MAY CUT power lines up to 8000 VOLTS.

Firefighters must be familiar with the hazards of electrical currents. The LOWEST that could be lethal to a well grounded person is 110 VOLTS.

There is little danger to firefighters directing hose streams on wires of LESS THAN 600 VOLTS to ground from any distance likely to be met under conditions of ordinary firefighting.

When cutting electrical service lines while the wires are overhead, the firefighter should, whenever possible, CUT THE LOOPS attached to the service head.

When an electrical wire has been pulled apart with NO CURRENT flowing, usually the break will have a CUP effect on one end and a CONE effect on the other end.

When an electrical wire has been cut with NO CURRENT flowing, usually it will have TWO WEDGE LIKE ENDS.

While an electrical wire IS CARRYING CURRENT during the break, it usually will have BOTH ENDS FUSED AND DISTORTED.

While fighting electrical fires, it must be remembered that water under high pressure becomes a GOOD CONDUCTOR.

If current doubles, heat will be FOUR TIMES AS GREAT.

RUBBER BOOTS should not be relied upon to supply the necessary resistance to ground, to prevent completion of electrical circuit through the body. (boots may contain carbon black).

ELECTRICAL GROUNDING: a connection between a conductive body and the earth that deletes the difference in potential between the article and ground.

There is a definite relationship between the LENGTH OF TIME (exposure) to electrical shock and the effects of the shock.

ELECTRIC TRANSFORMER: apparatus that converts and delivers electricity received at a certain voltage to electricity of a different voltage.

A shaft, within a building, is best defined as an enclosed space connecting a series of two or more openings in successive floors.

The property of matter by reason of which a body tends to resume its original shape when changed by an external force is termed as elasticity.

The force which holds together bodies of the same kind is termed as adhesion.

The term which is applied to the distance from the center of a circle to its boundary is known as the radius.

Automatic sprinkler systems are usually set in operation by heat.

Lubricant prevents rubbing surfaces from becoming hot because it forms a smooth film between the two surfaces, which prevents them from coming into contact.

A "spot weld" is when parts are held together in different places to hold them in place.

Pavement may crack during hot weather because heat expands.

Rust on tools and equipment is the result of oxidation.

TOOLS AND EQUIPMENT

All purpose tools may be used for a great variety of purposes.

Special tools are suitable only for a particular purpose.

AXE HANDLES are usually made of wood rather than steel because the wood tends to cushion the impact.

PICKHEAD AXE: a hand axe with a blade and a pick, also called FIREMAN'S AXE.

FLATHEAD AXE: is a single bitted axe with a flattened head.

PIKE POLE: is a wooden or fiberglass pole that has a metal point with hook. (usually used to pull down plaster from ceilings).

For opening of ceilings from below, use PIKE POLE or PLASTER HOOKS.

A CARBIDE-TIPPED CHAIN SAW is designed to cut:
1. Clay brick.
2. Asphalt.
3. Composition roofing material.

When using a SAWZALL the SHOE should be held FLAT against the material to be cut.

HOSE CLAMPS are used for shutting down the water from charged lines. The hose clamp is placed a minimum of 25 feet behind the apparatus and approximately 6 feet behind the coupling on the hydrant side.

Besides its use for opening doors that swing inward, the DETROIT DOOR OPENER is effective for the use as an emergency hose clamp.

FOUR-WAY VALVE: used on a hydrant primarily to permit the Fire Engineer to change from direct hydrant stream to pump stream without shutting of the water.

HALYARD: the rope used to extend an extension ladder.

MANILA ROPE is the standard for Fire Department rope.

COTTON FIBER ROPE: tensile strength is less than nylon fiber, sisal fiber, and manila fiber.

If there is a knot in a rope the rope will be WEAKENED at that point.

A temporary method of securing an object with a rope so that it can be easily undone is called a HITCH.

EYE SPLICE loop spliced at end of rope.
BOWLINE KNOT: will not slip or tighten under tension and it is easily untied. multipurpose knot.

The BEST method for holding the LIFE NET is with both palms facing the firefighter and thumbs alongside the first finger.

SCBA: Self-Contained Breathing Apparatus have no maximum time limit for changing air bottles as long as the pressure remains sufficient.

Air pressure in demand type breathing apparatus is 2000 PSI TO 2300 PSI.

Standard compressed air cylinder for breathing apparatus is rated for 30 MINUTES.

Standard compressed air cylinders (steel) are regulated by INTERSTATE COMMERCE COMMISSION (ICC).

Firefighters MAY NOT BE ABLE to ware breathing apparatus while responding to an incident, because the seat belts or safety tailboard straps may not work while wearing breathing apparatus.

Self Contained Underwater Breathing Apparatus (SCUBA) contain a mixture of OXYGEN AND NITROGEN.

SEQUENCE FOR USING STRAPS ON BREATHING APPARATUS FACE MASK:
    1. Neck.
    2. Side.
    3. Top.

Portable Oxyacetylene torch (cutting), Oxygen is set at 35 PSI.

PITOMETER: used to determine the flow capacity of fire hydrants.

PYROMETER: used for measuring high temperatures.

DECIBELS: used in measuring amounts of sound.

GALVANOMETER: used for measuring current-strength or potential difference.

GALVANOMETRY: is the science, art or process of measuring electric currents.

GALVANOMETER: is activated by electricity.

The HURST POWER TOOL uses a GASOLINE ENGINE.

MECHANICAL INFORMATION

WORN brake lining will cause squeaky brakes.

Unequal adjustment of brakes creates danger of SKIDDING.

BRAKE FADING on fire apparatus is caused by the brake drums and lining getting hot.

Brake lining on fire apparatus will SQUEAL if foreign material gets embedded in it.

EXCESSIVE HEAT is the cause of brake fade, which results in poor braking due to the rapid wearing away of the lining.

ORDINARY BRAKE FLUID is not used in the disk air brake system because it will break down too fast.

At 80 degrees F batteries are fully charged with a specific gravity reading of 1.280 on the hydrometer.

BATTERY ELECTROLYTE is sulfuric acid and water. A discharged battery has a LOWER specific gravity.

Gases released from storage batteries are HYDROGEN and OXYGEN.

Large amounts of water added to battery would indicate DEFECTIVE voltage regulator.

CURRENT FLOW in storage battery is drawn from positive to negative.

ELECTROLYTE is the liquid in a storage battery.

NEUTRALIZE battery acid with BICARBONATE OF SODA.

If a lead acid battery is discharging, the electrolyte will become WEAKER.

In a fully charged battery, the ELECTROLYTE is 1 1/4 TIMES AS HEAVY as pure water.

SULFURIC ACID is the acid used in lead acid storage batteries.

While replacing a battery, if you have a question as to the correct pole to ground, connect the battery and check the AMMETER.

A high charging rate is indicated by high battery water CONSUMPTION.

A lead acid storage battery with six cells connected in series will produce 12 VOLTS.

Keeping the plates covered in a battery is the MOST IMPORTANT thing that you can do to prolong the life of the battery.

The VOLTAGE in batteries can be increased by increasing the number of CELLS.

Operating fire apparatus electrical accessories with the engine OFF is a severe DRAIN on the battery.

Current from a storage battery is drawn from POSITIVE to NEGATIVE.

While a storage battery is recharged it liberates HYDROGEN.

When removing a battery, REMOVE THE GROUNDED TERMINAL FIRST.

High battery water CONSUMPTION usually indicates a high charging rate.

A HYDROMETER has a scale calibrated to read SPECIFIC GRAVITY.

Electrolyte of FULLY CHARGED BATTERY weighs 1 1/4 times as much as pure water.

Per-cent of BATTERY CHARGE to specific gravity:
1.265 to 1.280 = 100%
1.225 = 75%
1.200 = 55%
1.190 = 50%
1.155 = 25%

Fully charged battery will LOSE 10% of its power at 0 degrees F.

OVERCHARGING is the most common cause of battery damage.

The PRIMARY function of the CURRENT REGULATOR, as distinct from the voltage regulator, is to prevent the generator from charging beyond maximum-rated output.

The AMMETER on a fire apparatus shows the flow of the electric current to and from the storage battery.

If AMMETER needle is FLUCTUATING rapidly, the most likely cause is a faulty REGULATOR.

The PRIMARY function of the cutout relay in a GENERATOR is to keep the battery from discharging while the engine is idling.

The electrical system of an ALTERNATOR is prevented from reversing its flow of electricity from the battery by the use of a RECTIFIER.

FUSES and CIRCUIT BREAKERS protect against amperage overload.

PRIMARY function of voltage regulator is to keep battery from overcharging.

IGNITION SWITCH completes connection between battery and coil. (primary lead).

IGNITION COIL increases amperage and decreases voltage.

MAGNETO generates electric current.

CAPACITOR is part of the distributor.

CONDENSER is used in the ignition system.

DWELL ANGLE is the number of degrees a cam rotates while ignition points are closed.

PISTON: the part in a cylinder that moves up and down.

COMPRESSION GAUGE is used to check the cylinder pressure.

Cam follower is the VALVE LIFTER.

COIL: the device used to step-up low voltage to high voltage needed at the spark plugs.

DISTRIBUTOR: delivers the spark directly to the spark plugs.

CARBURETOR: is where gasoline vapor and air are mixed to the correct proportion.

Oil DILUTION is prevented in the crankcase by ventilation.

The OIL GAUGE on the dashboard indicates oil pressure within the engine.

As far as VISCOSITY, the LOWER S.A.E. rating is, the easier the oil will flow at lower temperatures.

RADIATOR pressure cap is used to raise the boiling point of the water in the radiator.

THERMOSTAT: usually operates with a bimetallic strip.

The primary purpose of a THERMOSTAT is to shut off the water circulation between the radiator and the engine, when the engine is cold.

SPEEDOMETER and TACHOMETER both use permanent magnets.

Excessive wear on the outside edge of a front tire is likely to be caused by too much CAMBER.

POSITIVE CASTER will be caused by tipping the top of the kingpin backward from the vertical position.

TOO LITTLE free movement in the clutch pedal, clutch may not engage.

TOO MUCH free movement in the clutch pedal will result in failure of the clutch to disengage properly.

Riding the clutch INCREASE the possibility of damaging the throw out bearings.

Thermostat has a BI-METALLIC strip.

Frequent and regular washing, waxing, and polishing will LENGTHEN the life of the painted finish and bright metal trim. (warm or cold water).

IDEAL engine operating temperature is just below the boiling point of the liquid being used.

RING GEAR is located in the standard differential.

A fire apparatus can go FASTER forward than in reverse because the gear ration is almost as high as low gear.

The TRANSMISSION is responsible for allowing the change in gear ratio between the engine and the rear wheels of fire apparatus.

Higher COMPRESSION ratio will give greater power at all speeds.

CAM FOLLOWER is a valve lifter.

CAM SHAFT operates the valve push rods.

Rotative force developed by engine is TORQUE.

The MAIN PURPOSE of a fire department pumper is to provide adequate pressure for fire streams.

To prevent engine damage while driving downhill, the speeds should be not more than 200 to 300 RPM above rated speed.

Oil gage needle FLUTTERING may indicate low oil.

Oil filters REMOVE sludge.

CHOKE: To restrict flow of a fluid.

EMULSIFIED OIL is caused by water vapor from burning fuel.

Racing gasoline engines prior to turning off ignition will DILUTE crankcase oil.

Water formed by burning fuel causes CORROSION of internal engine components.

Main effect of DETERGENTS added to lubricants is to hold contaminants in suspension.

CHOKE VALVE is for rich fuel-air mixture. (shuts off air intake).

Engine missing at high speeds is caused by partly stopped-up FUEL LINE.

AIR CLEANER will act as a flame arrester in case of engine backfire through the carburetor.

Perfect gasoline to air mixture = 15 PARTS AIR to 1 PART GASOLINE, at idle = 11 parts air to 1 part gasoline.

The device which gasoline vapor and air are mixed in proper proportion is known as the CARBURETOR.

The GREATEST SOURCE of engine oil contamination is normally from unburned fuel.

In an internal combustion engine the DETONATION is the result of the secondary ignition of the fuel charge after the regular spark occurs.

The HIGHER the octane rating of a fuel, the higher its ability to resist detonation.

Internal combustion engines having CONSTANT VOLUME combustion are usually those utilizing gasoline and air.

CHOKE DAMP: mixture of gases causing choking.

Apparatus engines should be checked for proper oil level with the ENGINE OFF, and approximately 30 minutes after shutting off.

TOO RICH fuel causes:
    1. Loping-sluggish engine.
    2. Irregular running engine.
    3. Engine overheating.
    4. Black smoke.
    5. Dirty air cleaner.

Engine cutting is usually CARBURETOR problems, engine cutting at high speeds is caused by partly stopped-up gas line.

When an engine will not start even though the starter turns over, the trouble is most likely the FUEL SUPPLY.

TOO LEAN of carburetor fuel mixture causes:
1. Motor popping back into manifold and carburetor.
2. Engine too slow or stops when accelerated suddenly.
3. Noticeable loss of power.

The PURPOSE of ventilating the crankcase is to prevent oil dilution.

Power developed by GASOLINE ENGINES decreases 3 1/2 % per 1000 foot altitude above sea level.

Power developed by DIESEL ENGINES decreases 3% per 1000 foot altitude above sea level. (normally aspirated diesel engines).

TURBOCHARGED DIESEL ENGINES do not have a power loss until altitudes in excess of 4000 feet and then only 2% per 1000 foot altitude above 4000 foot altitude level.

The PRINCIPAL advantage of obtaining a higher compression ratio in an fire apparatus engine is the engine will have greater power at all speeds.

Perfect GASOLINE TO AIR MIXTURE is 15 parts air to 1 part gasoline, at idle it is 11 parts air to 1 part gasoline.

Regulate speed and power in diesel engines by AMOUNT of fuel in the cylinders.

The quickest and safest way to find a spark plug not firing is to SHORT the plug to the engine with a wooden handled screw driver.

A partly stopped up fuel line may be indicated by the engine MISSING at high speeds.

VAPOR LOCK can be treated by moving the fuel line away from the exhaust manifold.

Ignition quality of diesel fuel is measured by OCTANE.

DIESEL engine detonation is usually EARLY fuel injection timing, when related to the injection timing.

COMPRESSION IGNITION is used in diesel engines.

To change engine speed or power in diesel engines you must VARY THE AMOUNT OF FUEL, the air entering always remains the same.

In most 4 STROKE engines 2 revolutions of the flywheel are completed with each cylinder cycle.

In 6 CYLINDER 4 CYCLE gas engines the breaker cam and distributor rotor will rotate at 1/2 the crankshaft speed.

In a gas engine that one explosion takes place in each cylinder once every two revolutions of a shaft is called a 4 CYCLE ENGINE.

LUBRICANTS prevent rubbing surfaces from becoming hot because of the smooth film that it creates between the surfaces, which prevent them from coming into contact.

If the OIL PRESSURE GAGE drops to 0 the FIRST thing that an apparatus driver should do is STOP the engine to prevent damage to the bearings.

To PREVENT oil dilution fire apparatus crankcases are ventilated.

The HIGHER the SAE number of engine oil = the higher the viscosity.

DETERGENTS in lubricants HOLD CONTAMINANTS IN SUSPENSION.

Engine overheating is usually 1 MAJOR PROBLEM rather than several minor problems.

TIRE WEAR:
   1. Under inflated   = both edges wear
   2. Overinflated   = center of tread wear
   3. Out of line   = wear on one edge
   4. Improper balance = flat spots, cupping
   5. Heavy braking   = flat spots, abrasions
   6. Toe in or out   = feathering
   7. Fast cornering   = feathering
   8. Camber   = tire wear on outside

Tire UNBALANCE causes shimmy.

CAMBER is the inward inclination of the front wheels at the bottom. Wear is on the outer edge. (refers to wheel adjustment).

The MOST important type of friction in the control of an apparatus is between the road surface and the tires.

A bumpy road REDUCES available friction between the road surface and the tires.

Excessive wear of the outer edge of the tread of the front tires is most likely the result of too much CAMBER.

CAMBER refers to the adjustment of the wheels of the vehicle.

In wheel alignment, the amount of TOE-IN is adjustable by changing the length of the tie rod.

The TILLERMAN on a fire apparatus is for controlling the rear wheels.

Driving habits of operators are the major factors for obtaining maximum FUEL ECONOMY from vehicles.

HIGHEST MANIFOLD VACUUM when operating an engine under its own power is obtained oat INTERMEDIATE SPEED, with the throttle partially opened.

On HEAVY APPARATUS the greatest driving torque reaction is transmitted through the drive train and is absorbed by the REAR SPRINGS.

IGNITION COIL: takes the current from the battery and generator and delivers it to the spark plugs with INCREASED AMPERAGE and DECREASED VOLTAGE.

ROUND WIRE FEELER GAUGE: is the most accurate tool to use to set the proper gap on a spark plug.

Measure the COMPRESSION in an internal combustion engine at the SPARK PLUG HOLE.

PRE-IGNITION, with ignition switch off is most likely caused by an overheated engine.

MAXIMUM TORQUE is normally obtained from an engine at speeds BELOW the engine speeds at which maximum horsepower is produced.

ONE HORSEPOWER is the power required to lift 550 LBS 1 foot in 1 second.(33,000 LBS 1' in 1 minute)

CAM SHAFT: operates the valve push rods.

The RING GEAR is located in the STANDARD DIFFERENTIAL.

A 4-stroke cycle gas engine with a crankcase gear of 24 teeth will have a TIMING GEAR on the camshaft of 48 TEETH.

SPUR GEARS in conventional transmissions have teeth that appear to run straight across the wheel.

The BEVEL GEARS most common use is to transmit power from two shafts whose axes intersect at right angles.

Fire Department storage batteries fluid = DILUTED SULFURIC ACID.

AMMETER: measures the amperage of current.

ELECTROLYTE: used in a wet cell battery, is a chemical compound which will decompose when an electric current is passed through it.

Generator, transformer and voltage regulator are used to GENERATE or CONTROL electrical currents.

OVERCHARGING THE BATTERY is the most frequent cause of battery failure in the Fire Service.

RECTIFIER works with the alternator.

CURRENT REGULATOR protects the generator.

VOLTAGE REGULATOR protects the battery.

Ideal operating engine temperature for liquid cooled engines is JUST BELOW THE BOILING POINT of the liquid being used.

AMBIENT TEMPERATURE of vehicle engine = temperature inside the engine compartment.

MECHANICAL PRINCIPLES

PULLEYS:

1. The more pulleys the easier it is to pull or lift an object.
2. The more pulleys involved the greater distance you must pull, but it is still easier to lift an object.
3. The thinner a windlass, the easier it is to turn.
4. In two different sets of pulleys, if the wheels are connected by a shaft, and the two wheels on one pulley are the same as the other two that they are connected to, then they both turn at the same speed.

BELTS:

1. Always determine in which direction one of the wheels in a diagram is turning as the belt will be going in the same direction. Also you can determine the direction of the belt and the wheel direction will be the same.
2. Wheels under a belt that is not twisted all turn in the same direction. Those on the outside of the same belt would turn in the opposite direction of those on the inside.

WHEELS:

1. If wheels are a different size on the same vehicle, then the smaller wheel will turn faster.
2. When wheels of different sizes are joined together by belts, the smallest wheel turns fastest, the largest wheel slowest.
3. When two gears of different sizes are locked together, the smaller gear turns faster than the shaft connected to the larger gear.

TURNING/DIRECTIONS:

1. The faster an object whirls around, the more it will pull from the center of rotation.

2. If a car, or tractor or objects are turning, then the inside wheels or objects will turn less distance and more slowly than the outside ones.

3. When a car skids, its speed increases momentarily to the outside when turning.

CENTER OF GRAVITY: (referring to the point which weight is evenly distributed)

1. A solid object with a space drilled out will rest on the section that is solid.

2. The higher a vehicle is packed with materials the easier it will turn over when on an incline.

VOLUMES AND AREAS OF SOLID OBJECTS:

1. If several solid figures have the same width and height but different shapes then their weight and volumes are different. The lowest weight or least volume is a solid of triangular shape. Then comes a cylindrical solid (circular in shape) and then a cube (square shape)

2. Objects (cars) placed or parked parallel to each other and perpendicular to the side occupy less space.

MECHANICAL QUESTIONS
QUESTIONS: PULLEYS, GEARS, AND LEVERS

1. Of the two block & tackles below, which one will lift a load more easily?

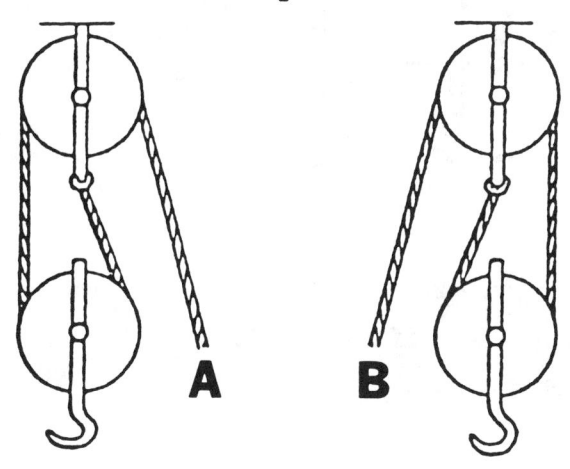

A. Drawing #1 B. Drawing #2 C. No difference.

ANSWER = C

2. Cylinder #1 and cylinder #2 turn in oppsite directions at the same time. If #1 turns in the direction of the arrow, in what direction will the pulley hook travel?

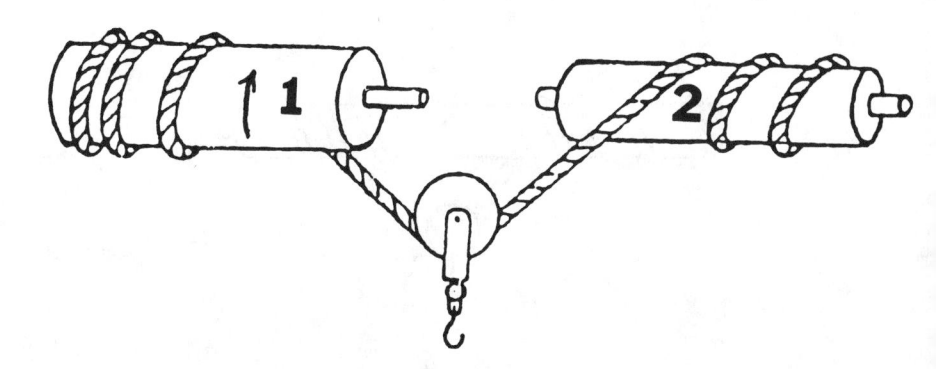

A. In an upward direction.
B. In a downward direction.
C. It will not travel up or down.

ANSWER = B

3. Of the two drawings below, which winch will be able to lift a heavier load?

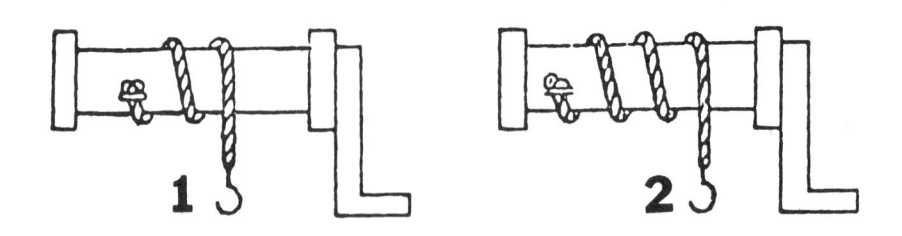

1

2

A. Drawing #1.
B. Drawing #2.
C. They will lift equal loads.

ANSWER = C

4. In the drawing below, if the cylinder is rotated in the direction of the arrow, in what direction will the pulley hook go?

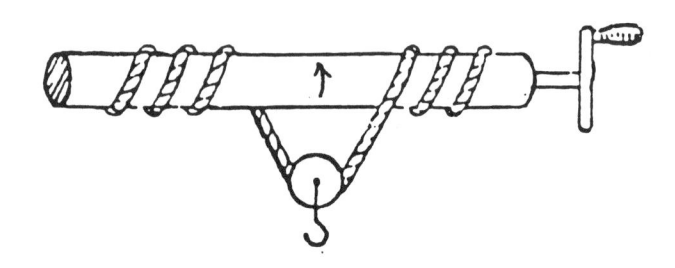

A. Downward.
B. Upward.
C. Neither direction.

ANSWER = C

5. If the large cylinder turns in the direction of the arrow, in what direction will the pulley hook go?

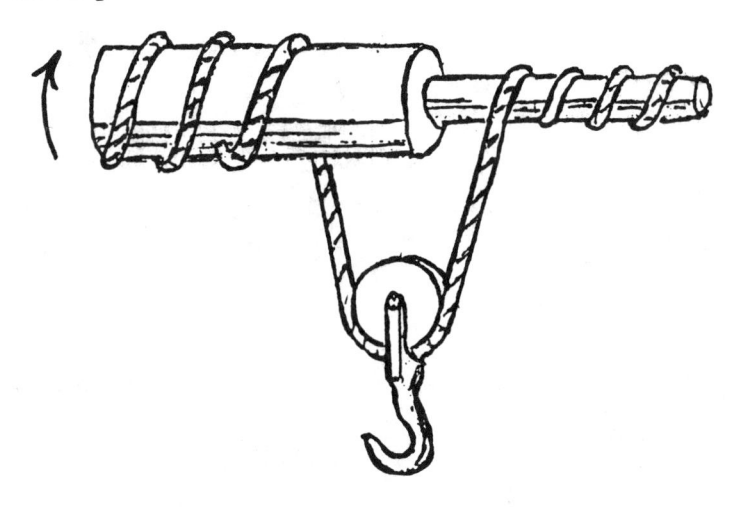

   A. Upward.
   B. Downward.
   C. Neither direction.

   ANSWER = B

6. In which direction must the larger gear rotate in order to make the "C" rotate?

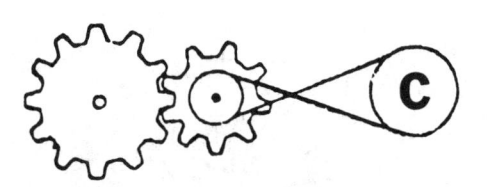

   A. Counterclockwise.
   B. Clockwise.
   C. Either counterclockwise or clockwise.

   ANSWER = C

7. If gear "X" is rotated in the direction of the arrow, which direction will gear "Y" rotate?

A. Clockwise.
B. Counterclockwise.
C. Either clockwise or counterclockwise.

ANSWER = A

8. In the drawing below, which direction can the sprocket rotate?

A. Counterclockwise.
B. Clockwise.
C. In both directions.

ANSWER = A

9. Which direction must the inside gear be rotated in order to rotate the ouitside gear?

A. Counterclockwise.
B. Clockwise.
C. In either direction.

ANSWER = C

10. In the drawing below:

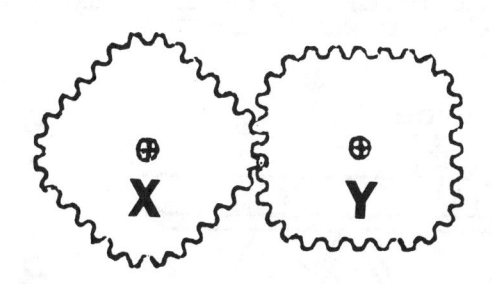

A. For each turn of gear "X", gear "Y" will make a complete rotation.
B. Only gear "X" can be the driving gear.
C. Wheel "Y" will move more slowly than Wheel "X".
D. None of the above.

ANSWER = C

11. In the drawing below:

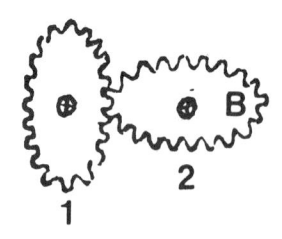

A. When gear #1 touches gear #2 at the "B" position, gear #2 will be moving slower than gear #1.
B. The gears will not turn.
C. Gear #2 can be the driving gear.
D. None of the above.

ANSWER = A

12. In the drawing below, when a force is exerted at the location of the #1, on the teeterboard drawing below, in which direction will end #2 go?

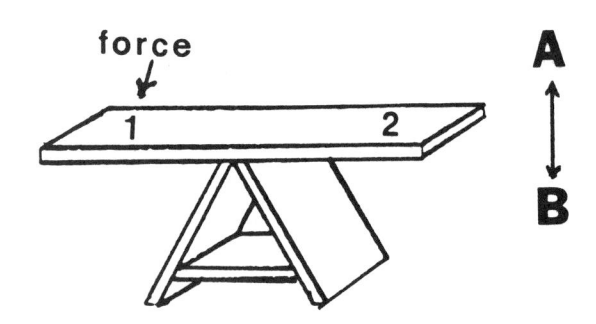

A. Up.
B. Down.
C. Neither direction.

ANSWER = A

126

13. In the drawings below, which load will be the easier to lift when exerting pressure at the location of the arrows?

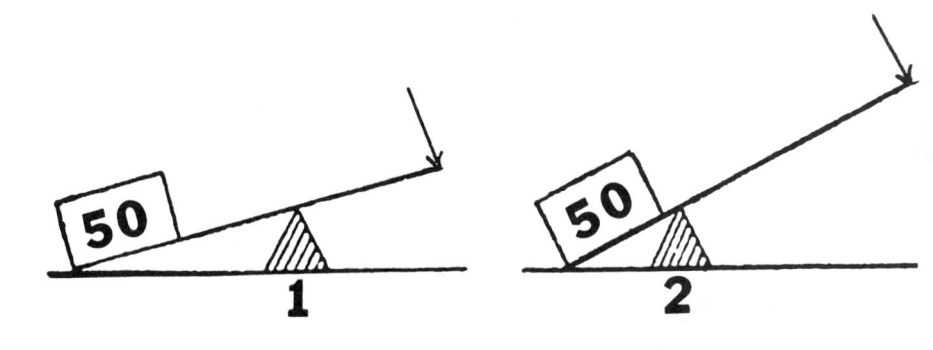

A. Drawing #1.
B. Drawing #2.
C. It will not make any difference.

ANSWER = B

14. In the drawing below, which block will be the most difficult to tip over?

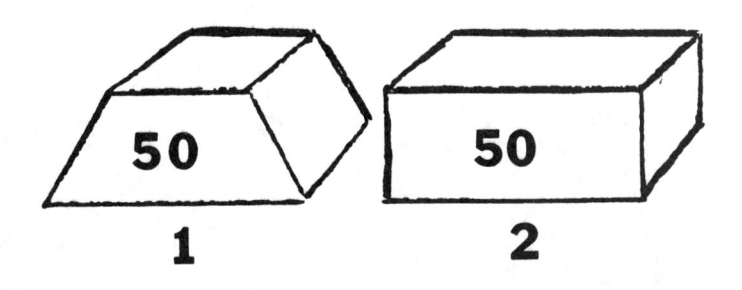

A. #1.
B. #2.
C. There would be no difference.

ANSWER = C

15. In the drawing below, which chain has the greatest stress on it?

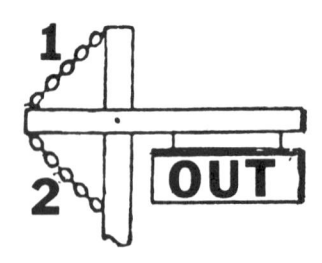

A. Chain #1.
B. Chain #2.
C. They have equal stress.

ANSWER = C

16. In the drawing below, if #1 and #2 are of equal size and weight, and the pulley at the top will turn freely, which one of the pulleys will tend to roll down the ramp?

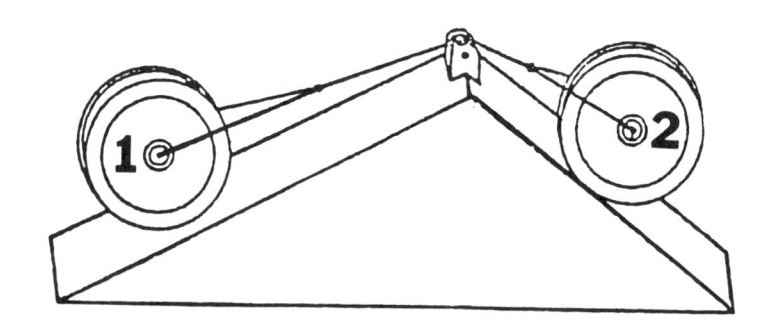

A. #1.
B. #2.
C. It will not make any difference.

ANSWER = B

17. In the drawing below, which ladder is the LEAST likely to tilt over?

1

2

A. Ladder #1.
B. Ladder #2.
C. It does not make any difference.

ANSWER = A

18. In the drawing below, which teeter would be the most stable?

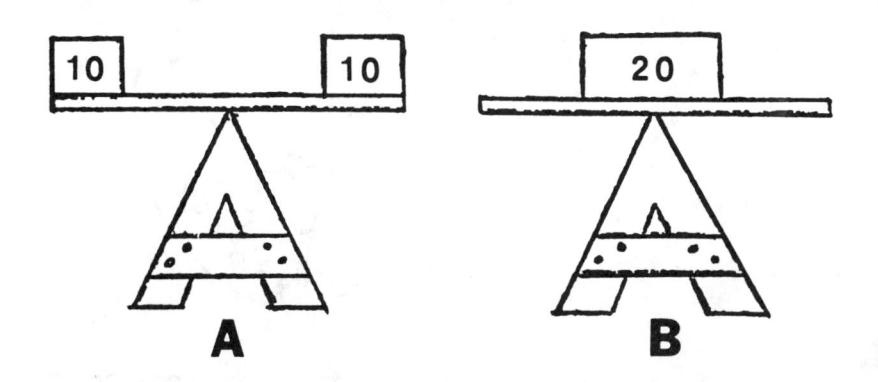

A

B

A. Teeter A.
B. Teeter B.
C. They are of equal stability.

ANSWER = B

19. In the drawing below, in order to SHORTEN the turnbuckle, in which direction should it be turned?

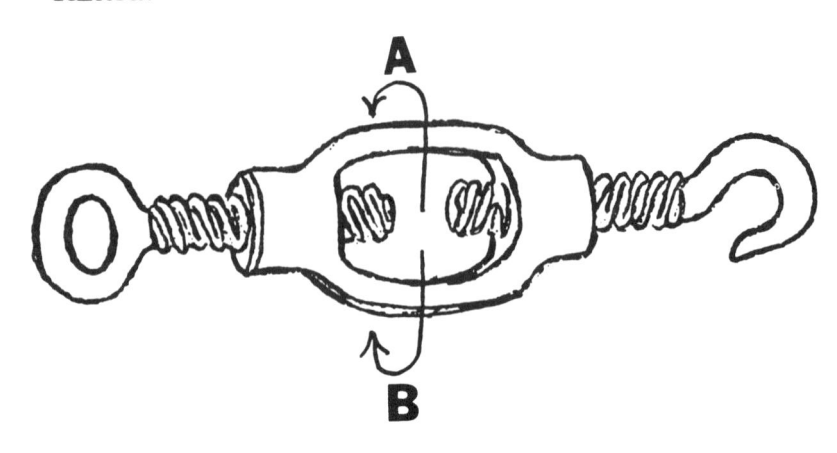

   A. Direction A.
   B. Direction B.
   C. It would not make any difference.

   ANSWER = A

20. In the drawing below:

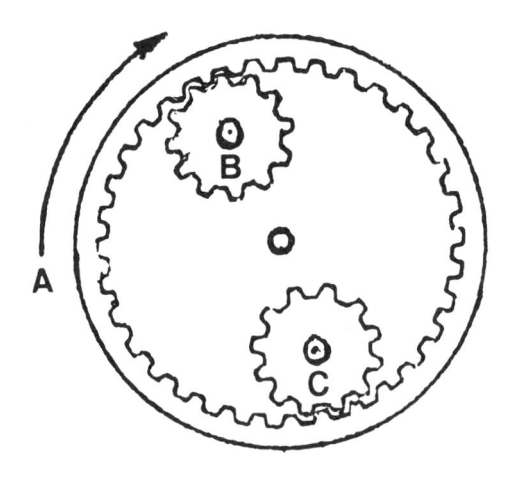

   A. Gear A and gear B will turn in opposite directions.
   B. Gear B and gear C will turn in opposite directions.
   C. Answers A and B are both correct.
   D. None of the above.

   ANSWER = D

21. Which block and tackle will lift the load faster?

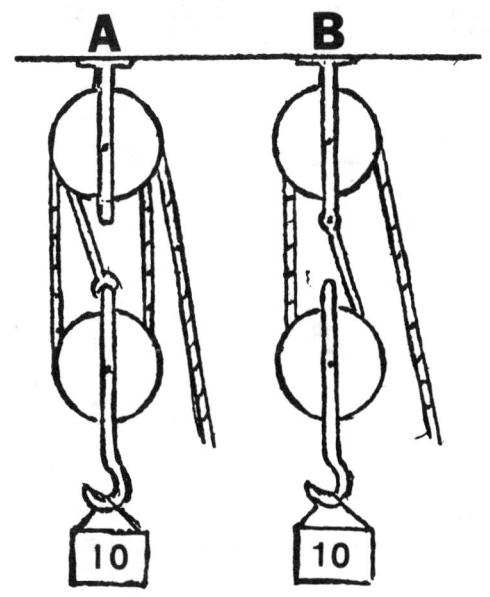

A. Block and Tackle A.
B. Block and Tackle B.
C. They will lift at the same rate.
ANSWER = A

22. Which winch will raise the weight more easily?

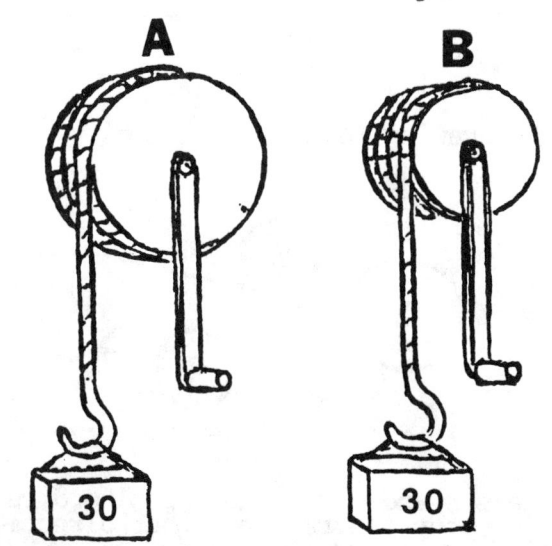

A. Winch A.
B. Winch B.
C. They will both lift equally.
ANSWER = A

23. In the drawing below, if wheel A and wheel B make a complete revolution every second, which notch moves the fastest?

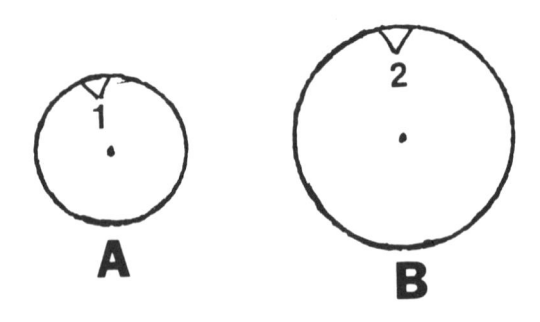

A. Notch #1.
B. Notch #2.
C. Bothe move at an equal rate of speed.

ANSWER = A

24. In the drawing below:

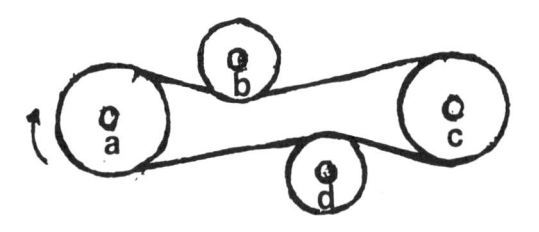

A. Wheels b and d turn in opposite directions.
B. This setup can run in either direction.
C. All four of the wheels will run at the same speed.
D. None of the above.

ANSWER = B

25. In the drawing below:

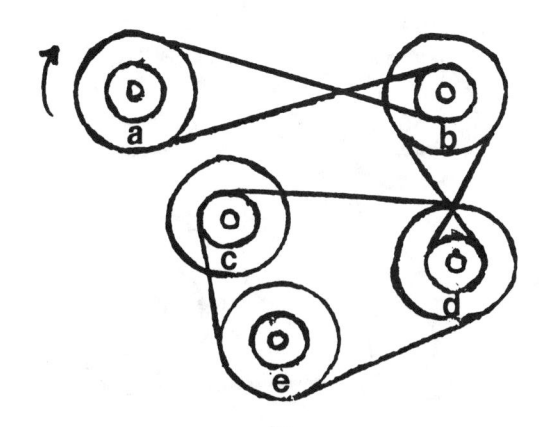

A. Wheels a and c turn in opposite directions.
B. Wheels a and d turn in opposite directions.
C. Wheel d turns faster than wheel a.
D. Wheel b turns faster than wheel e.

ANSWER = D

26. In the drawing below:

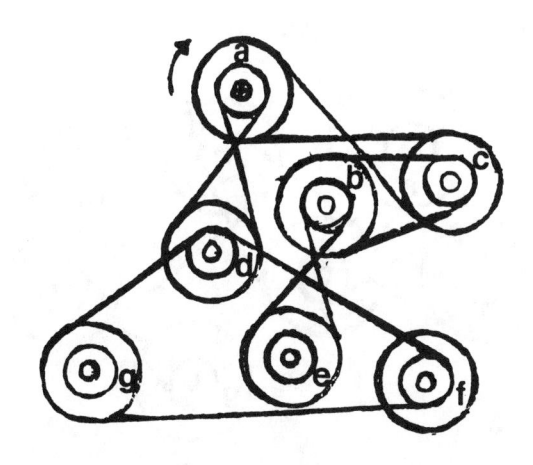

A. Wheels c and d turn in the same direction.
B. Wheels a and f turn at the same speed.
C. Wheel g turns faster than wheel b.
D. None of the above.

ANSWER = A

27. In the drawing below:

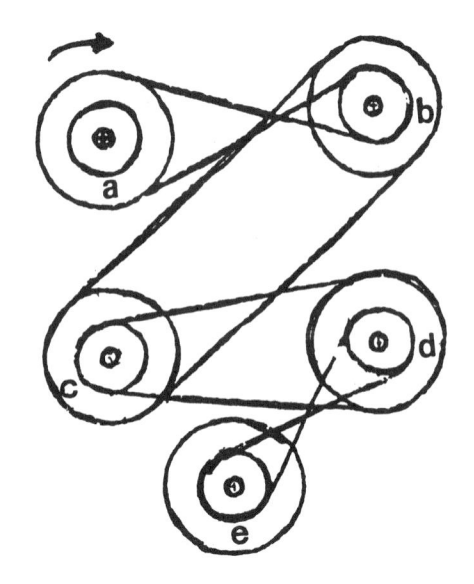

A. Wheels a and d turn in the same direction.
B. Wheel d turns faster than wheel b.
C. Wheel a turns faster than wheel c.
D. Wheels d and a turn at the same rate.

ANSWER = B

28. In the drawing below:

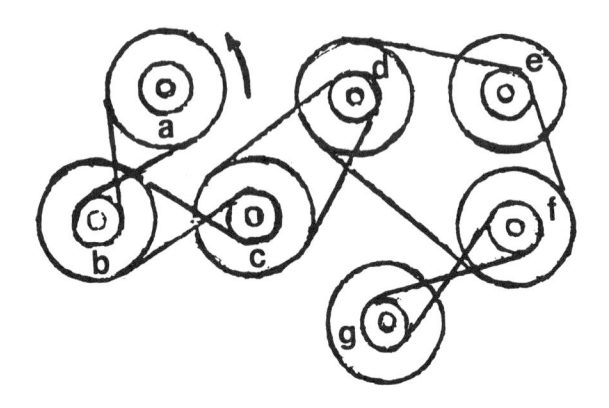

A. Wheels a and g turn in the same direction.
B. Wheels b and e turn in opposite directions.
C. Wheel c makes more revelutions than wheel g.
D. Wheel b makes more revelutions than wheel e.

ANSWER = B

29. In the drawing below, one complete turn of the drum crank will move the weight vertically upward a distance of?

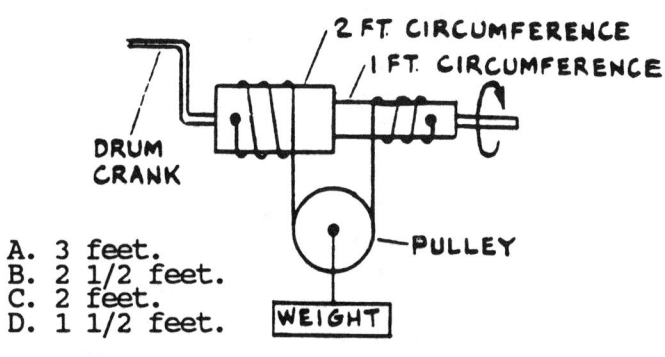

2 FT. CIRCUMFERENCE
1 FT. CIRCUMFERENCE
DRUM CRANK
PULLEY
WEIGHT

A. 3 feet.
B. 2 1/2 feet.
C. 2 feet.
D. 1 1/2 feet.

ANSWER = D

30. In the drawing below, the maximum weight (w) that can be lifted with a 75 pound pull as shown is?

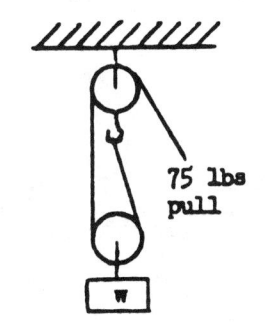

75 lbs pull

A. 50 pounds.
B. 75 pounds.
C. 100 pounds.
D. 150 pounds.

ANSWER = D

31. In the drawing below, which shaft will turn the slowest?

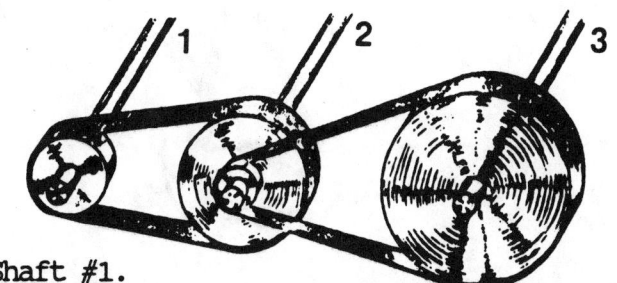

A. Shaft #1.
B. Shaft #2.
C. Shaft #3.
D. They will all rotate at the same speed.

ANSWER = C

32. In the drawing below, while lifting the load the required pull decreases:

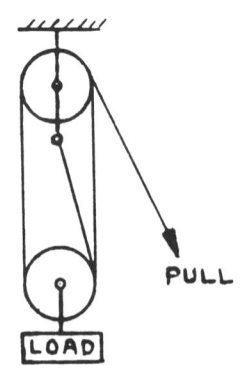

   A. As the distance between the pulley wheels decreases.
   B. When the rope size is increased.
   C. When the number of pulley are increased.
   D. If larger pulley wheels are used.

   ANSWER = A

33. In the drawing below, when the top pulley rotates in the direction of the arrow, in what direction will the lower pulley rotate?

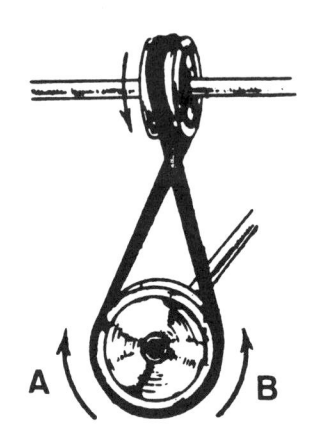

   A. Direction A.
   B. Direction B.
   C. In either direction.

   ANSWER = A

QUESTIONS 34, 35, AND REFER TO THE DRAWING BELOW. WHEELS "A" AND "B" HAVE THE SAME DIAMETER. WHEEL "C" HAS A DIAMETER 1/3 THE DIAMETER OF WHEEL "A". WHEEL "A" IS CONNECTED TO WHEEL "C" BY A BELT, AND WHEEL "B" IS FIXED TO WHEEL "C". WHEEL "A" ROTATES IN A CLOCKWISE DIRECTION AT A SPEED OF 10 RPM.

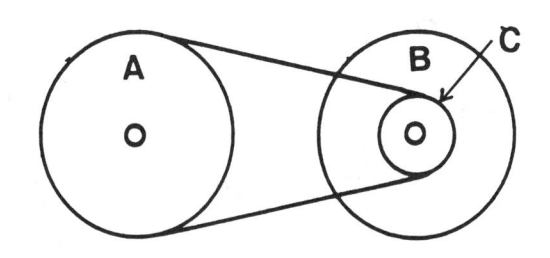

34. If wheel "A" rotates clockwise direction then:
A. Wheels "C" and "A" turn in opposite directions.
B. Wheels "B" and "A" turn in the same direction.
C. Wheels "C" and "B" turn in opposite directions.
D. Wheels "A" and "B" turn clockwise and "C" turns counterclockwise.

ANSWER = B

35. If wheel "A" turns at a rate of 10 rpm, the number of rpm made by wheel "C" is most nearly:
A. 90 rpm.
B. 60 rpm.
C. 30 rpm.
D. 3 rpm.

ANSWER = C

36. In the drawing below, if wheel #1 is turned in a clockwise direction, wheel #2 will?

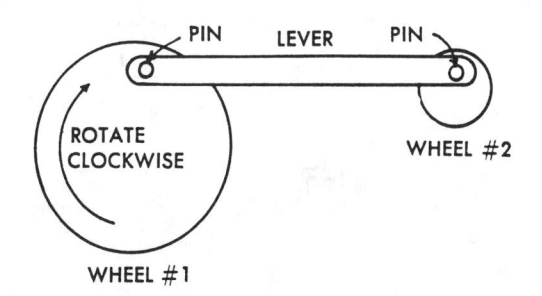

A. Move back and forth.
B. Continue to rotate in a clockwise rotation.
C. Rotate in a counterclockwise rotation.
D. Wheels #1 and #2 will jam and stop rotating.

ANSWER = D

37. Of the four following drawings, which one has the weights that will move?

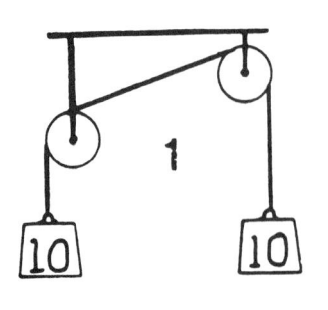

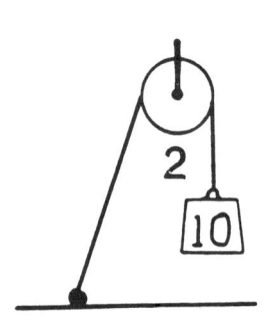

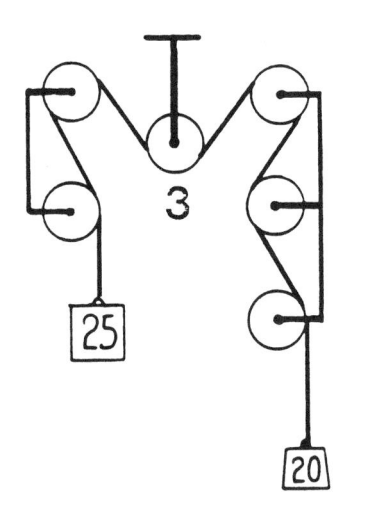

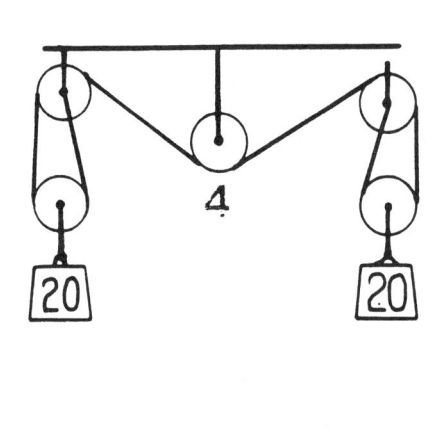

A. Drawing #1.　　　　　　　　B. Drawing #2.
C. Drawing #3.　　　　　　　　D. Drawing #4.

ANSWER = C

38. Of the following two drawings, which one has the weights that will move?

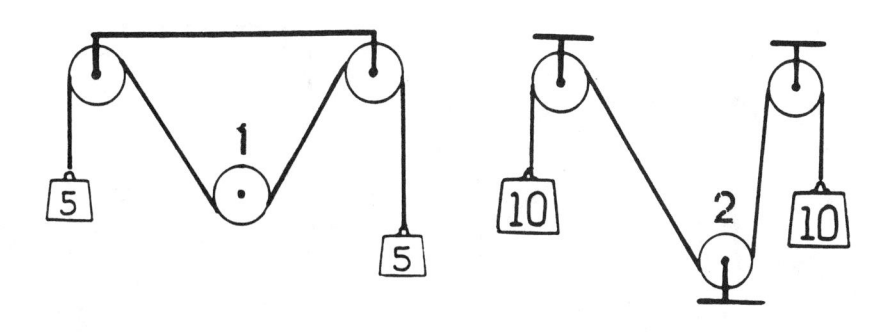

A. Drawing #1.   B. Drawing #2.

ANSWER = A

39. Of the two following drawings, which one has the weights that will move?

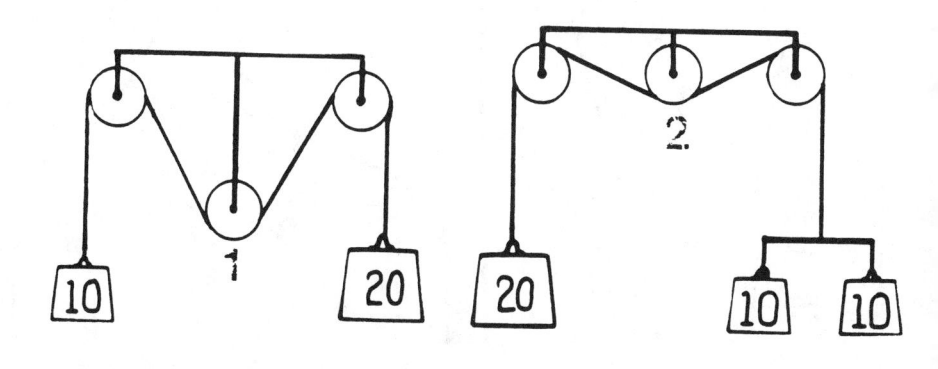

A. Drawing #1.   B. Drawing #2.
C. Both drawing #1 and #2.   D. Neither drawing.

ANSWER = A

40. Of the following four drawings, which one has the weights that will move?

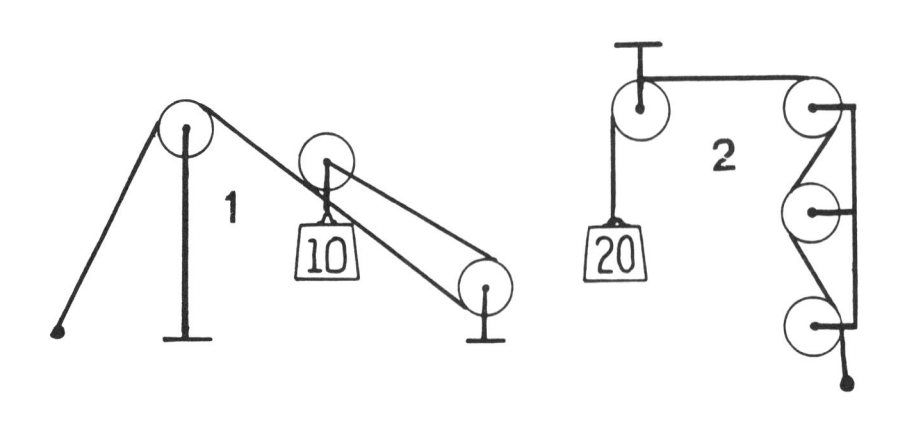

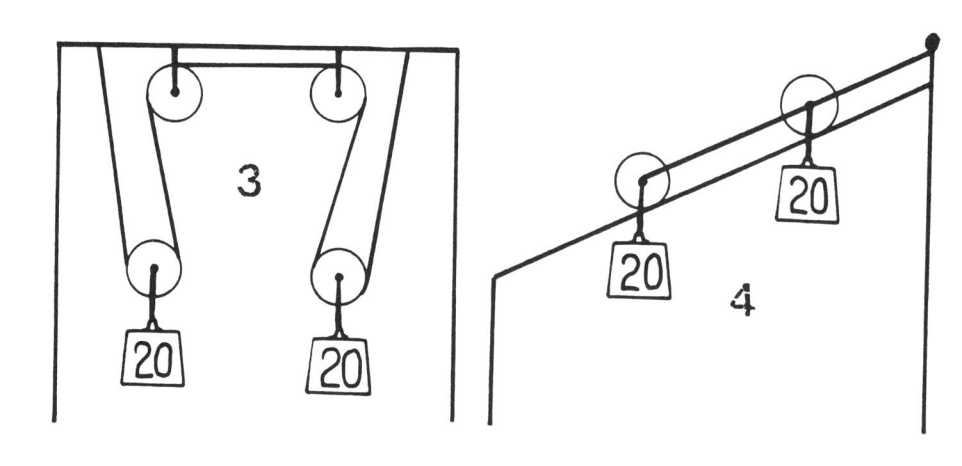

A. Drawing #1.
C. Drawing #3.

B. Drawing #2.
D. Drawing #4.

ANSWER = A

41. Of the two following drawings, which one has the weights that will move?

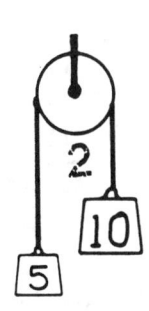

A. Drawing #1.
B. Drawing #2.
C. Both drawing #1 and #2.
D. Neither drawing.

ANSWER = B

42. Of the two following drawings, which one has the weights that will move?

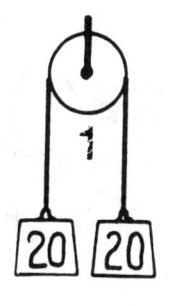

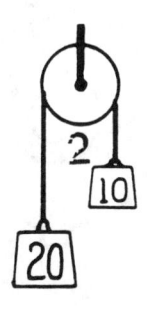

A. Drawing #1.
B. Drawing #2.
C. Both drawing #1 and #2.
D. Neither drawing.

ANSWER = D

43. In the drawing below, by crossing the pulley belt:

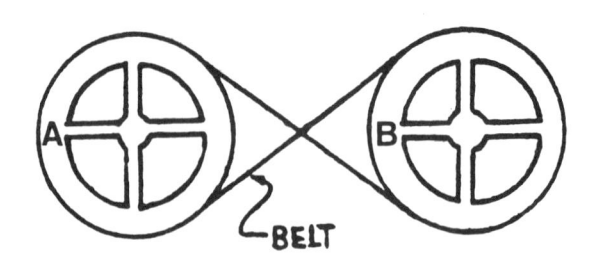

A. The slack will be taken up.
B. The wheels will turn in opposite directions.
C. The belt will last longer.
D. Wheel "A" will turn faster than wheel "B".

ANSWER = B

44. In the drawing below, the small gear will:

A. Decrease the speed of gears "A" and "B".
B. Increase the speed of gears "A" and "B".
C. Make gears "A" and "B" rotate in opposite directions.
D. Make gears "A" and "B" rotate in the same direction.

ANSWER = D

45. In the drawing below, pulley "A" rotates in the direction of the arrow, then pulley "B" will rotate:

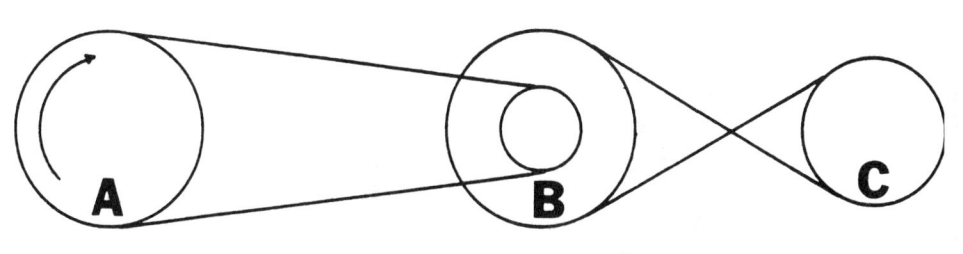

A. Slower than "A", in the opposite direction.
B. Faster than "A", in the opposite direction.
C. Slower than "A", in the same direction.
D. Faster than "A", in the same direction.

ANSWER = B

46. In the drawing below, if gear "a" rotates in the direction of the arrow, which gear will rotate the fastest?

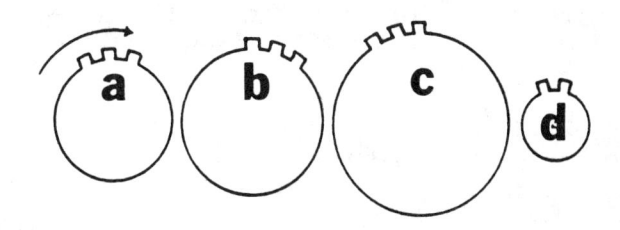

A. Gear "a"
B. Gear "b"
C. Gear "c"
D. Gear "d"

ANSWER = D

QUESTIONS 47, 48, 49, AND 50 ALL RELATE TO THE
DRAWING BELOW, WITH GEAR "A" THE DRIVING GEAR
ROTATING IN THE DIRECTION OF THE ARROW. GEARS "A"
AND "D" HAVE TWICE AS MANY TEETH AS GEAR "B", AND
GEAR "C" HAS FOUR TIMES AS MANY TEETH AS GEAR "B".

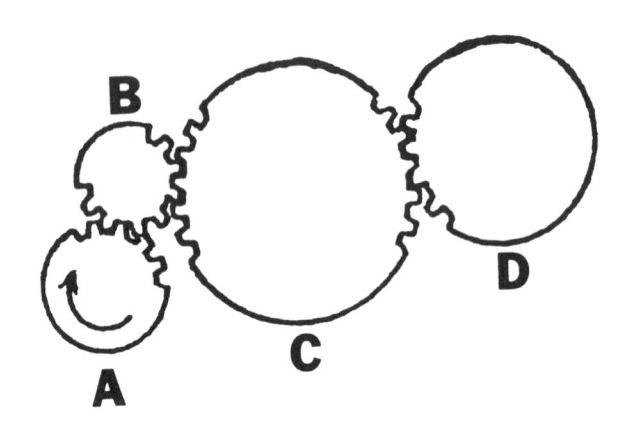

47. Which two gears turn in the same direction?
    A. Gears "A" and "B".
    B. Gears "B" and "C".
    C. Gears "C" and "D".
    D. Gears "B" and "D".
    ANSWER = D

48. Which two gears rotate at the same rate?
    A. Gears "A" and "C".
    B. Gears "A" and "D".
    C. Gears "B" and "C".
    D. Gears "B" and "D.
    ANSWER = B

49. If all of the teeth on gear "C" are knocked off,
    without influencing the teeth on gears "A", "B"
    and "D", then rotation would only take place
    with:

    A. Gear "C".
    B. Gear "D".
    C. Gears "A" and "B".
    D. Gears "A", "B", and "D".
    ANSWER = C

50. If gear "D" rotates at a rate of 100 rpm, then
    gear "B" rotates at a rate of:
    A. 10 rpm.
    B. 25 rpm.
    C. 50 rpm.
    D. None of the above.
    ANSWER = C

MECHANICAL QUESTIONS

1. The use of an absorbent material placed on a slippery surface, so as to reduce the chances slipping, the affect of an absorbent material is to increase:
   A. Force.
   B. Energy.
   C. Friction.
   D. Gravity.

   ANSWER = C

2. Room temperature is normally higher near the ceiling of a room than near the floor, because:
   A. Hot water pipes are near the ceiling.
   B. The warmer air is lighter than the cooler air.
   C. The air circulates better at floor level.
   D. Most openings are nearer the floor.

   ANSWER = B

3. Two weights of the same size but of unequal weights are dropped at the same time from the same height:
   A. The lightest weight will hit the ground first, because will be greater on the heavier weight.
   B. The heaviest weight will hit the ground first, because it weighs more.
   C. The two weights will hit the ground at the same time, because effect of gravity is the same on both weights.
   D. The two weights will hit the ground at the same time, because they are the same size.

   ANSWER = C

4. The reason it is not appropriate to increase the leverage of a wrench with the use of a pipe over the handle of the wrench, is the wrench:
   A. Will be more difficult to use.
   B. Will be more difficult to put on the nut.
   C. Will slip off the nut.
   D. Could break.

   ANSWER = D

5. The sudden closing of a nozzle on a fire hose that is discharging water under pressure may:
   A. Cause the hose to rupture.
   B. Cause the hose to flail.
   C. Cause the valve to jamb.
   D. Cause the valve to leak.

   ANSWER = A

145

6. When drilling through wood, you should clamp an additional piece of wood on the underside, to:
   A. Guide the drill bit.
   B. Drill through faster.
   C. Stabilize the drill bit.
   D. Prevent the wood from splintering.

   ANSWER = D

7. To help prevent steel beams from collapsing, during a fire, a layer of concrete may be applied to the steel beams. Why?
   A. Will cause the beams to be stronger during a fire.
   B. Insulates the beams.
   C. Reduces rust and corrosion.
   D. Will cause a chemical reaction during a fire.

   ANSWER = B

8. A fire pump is discharging 250 PSI through 200 feet of fire hose at ground level. When the nozzle is shut off, the pressure at the nozzle =
   A. The same as at the fire pump.
   B. More than the pressure at the fire pump.
   C. Less than the pressure at the fire pump.
   D. More or less than the pressure at the fire pump, depending on the type of fire pump used.

   ANSWER = A

9. Of the following, the most powerful and positive type of clutch is:
   A. Ring.
   B. Flush.
   C. Jaw.
   D. Cone.

   ANSWER = C

10. Of the following valves, the one that should be used either entirely open or fully closed, is:
    A. Globe.
    B. Gate.
    C. Pressure.
    D. reducing.

    ANSWER = B

11. The proper use of the tension pulley on a belt that connects two different size pulleys:
    A. Will not add to the life of the pulley.
    B. Usually positioned closest to the smaller pulley.
    C. Usually positioned closest to the larger pulley.
    D. Usually will be a "V" type pulley.

    ANSWER = B

12. In comparing gasoline engines to diesel engines:
   A. There is less soot produced by diesel engines.
   B. The use of oil additives is not recommended in gasoline engines.
   C. Diesel engines operate at higher temperatures.
   D. Gasoline engines operate at higher temperatures.

   ANSWER = C

13. It is considered poor practice to increase the leverage of a wrench by placing a pipe over the handle of the wrench, because:
   A. The wrench may break.
   B. The wrench may slip from the nut.
   C. It is harder to place the wrench on the nut.
   D. The wrench is more difficult to handle.

   ANSWER = A

14. The type of clutch which gives the most powerful and positive drive is:
   A. Jaw clutch.
   B. Ring clutch.
   C. Disk clutch.
   D. Cone clutch.

   ANSWER = A

15. A tension pulley properly used on a belt connecting two pulleys of different size:
   A. Will commonly be a crowned tension pulley.
   B. Is generally placed nearer the large pulley.
   C. Is generally placed nearer the small pulley.
   D. In no way adds to the belt life.

   ANSWER = C

16. Generally speaking, the practice of racing a car engine to warm it up, is:
   A. Good since repeated stalling of the engine and drain on the battery is avoided.
   B. Good since the engine becomes operational in the shortest period of time.
   C. Bad since proper lubrication is not established soon enough.
   D. Bad since too much fuel is used to get the engine warmed-up.

   ANSWER = C

17. When starting a vehicle equipped with a manual choke, on a cold day, it is best to pull the choke out, because it:
   A. Increases the amount of air in the carburetor.
   B. Reduces the amount of fuel entering the carburetor.
   C. Allows more fuel to enter the carburetor for a richer fuel mixture.
   D. Speeds up the supply of air and fuel to the engine.

   ANSWER = C

18. The best reason for lubricating moving parts of machines is to:
   A. Prevent the formation of rust.
   B. Reduce friction.
   C. Increase inertia.
   D. Reduce the accumulation of dirt on parts.

   ANSWER = B

19. The reason alcohol is added to the radiator of a vehicle in cold weather is because it:
   A. Lowers the freezing point of the mixture.
   B. Lowers the boiling point of the mixture.
   C. Raises the freezing point of the mixture.
   D. Raises the boiling point of the mixture.

   ANSWER = A

20. The water stream from a hoseline that is 100 feet in length has a reach farther than the water stream from a hoseline, from the same pump at the same engine pressure, that is 200 feet long, Why?
   A. The rise of temperature is greater in the longer hose length.
   B. Air resistance to the water stream is proportional to the length of hose.
   C. The time required for water to travel through the longer hose is greater.
   D. The loss due to friction is greater in the longer hose.

   ANSWER = D

21. Firemen usually lean forward when using a charged a charged hose line. What is the best reason for this?
   A. The fireman is more comfortable because of the cooling of the surrounding air at the tip.
   B. A backward force is developed which must be counteracted.
   C. The firemen are better protected.
   D. The firemen can see where the stream strikes better from this position.

   ANSWER = A

22. When water is traveling from a fire stream, and is directed at the roof of a three story building, at what point would the water be traveling at its greatest speed.
A. As it falls on the roof.
B. At its maximum height.
C. At a midway point between the ground and the roof.
D. As it leaves the hose nozzle.

ANSWER = D

23. The ammeter of a vehicle will indicate the flow of electrical current:
A. From the battery to the starter.
B. Outside the starting circuit.
C. To the lights.
D. To and from the battery.

ANSWER = D

24. If the temperature gauge in a vehicle indicates that the engine is staring to overheat:
A. Pour cold water in immediately.
B. Pour hot water in it immediately.
C. Pour in anti-freeze immediately.
D. Allow it to cool down.

ANSWER = D

25. A weight is to be supported from a brace by a chain of 30 links and a hook. Each link of the chain weighs 2 pounds and can support a weight of 1,000 pounds, and the hook weighs 10 pounds and can support a weight of 5,000 pounds, what is the maximum load that can be supported from the hook?
A. 25,000 pounds.
B. 5,000 pounds.
C. 1,000 pounds.
D. 930 pounds.

ANSWER = D

MATCHING TOOLS

MATCH THE PROPER TOOL FROM COLUMN #1 TO BE USED WITH THE PROPER FASTENING DEVICES IN COLUMN #2:

COLUMN #1

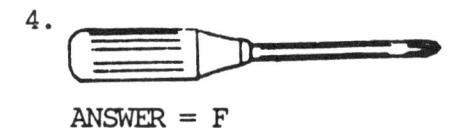

1.  ANSWER = E
2.  ANSWER = G
3. 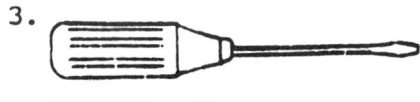 ANSWER = B
4. ANSWER = F
5.  ANSWER = D
6.  ANSWER = H
7. 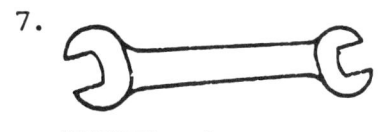 ANSWER = A
8.  ANSWER = C

COLUMN #2

A. 
B. 
C. 
D. 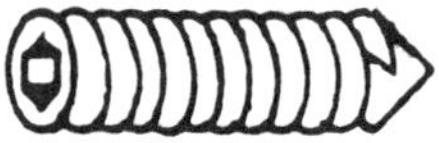
E.
F. 
G.
H.

MATCHING MECHANICAL PARTS

MATCH THE PROPER MECHANICAL PART FROM COLUMN #1 TO AN ASSOCIATED MECHANICAL PART IN COLUMN #2:

COLUMN #1   COLUMN #2

1. 
ANSWER = G

2. 
ANSWER = C

3. 
ANSWER = E

4. 
ANSWER = F

5. 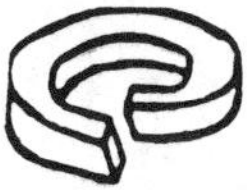
ANSWER = D

6. 
ANSWER = A

7. 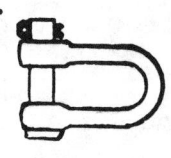
ANSWER = B

A.

B.

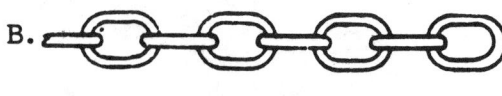

C.

D.

E.

F.

G.

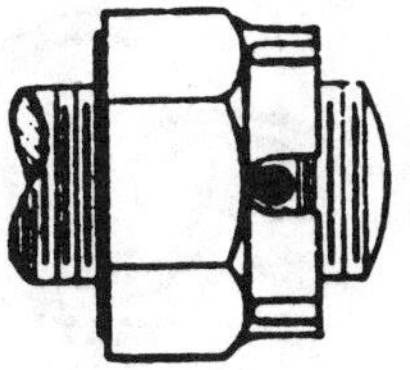

MATCHING KNOTS

COLUMN #1   COLUMN #2

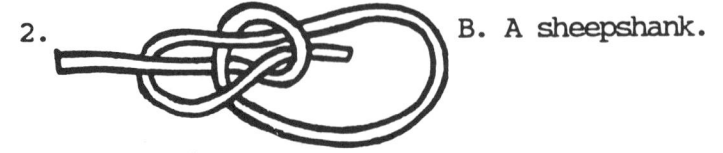

1. ANSWER = F

   A. A bowline.

2. ANSWER = A

   B. A sheepshank.

3. ANSWER = B

   C. A clove hitch.

4. ANSWER = E

   D. Two half hitches.

5. ANSWER = C

   E. A knot which may safely be used for joining two ropes of different sizes.

6. ANSWER = D

   F. A knot which may safely be used for joining two ropes of the same size, but is not safe for ropes of different sizes.

MEHANICAL OBJECT QUESTIONS

1. The object shown below is a:

    A. Ballpeen hammer.
    B. Framers hammer.
    C. Claw hammer.
    D. tack hammer.

    ANSWER = C

2. The object shown below is a:

    A. T-square.
    B. Depth gauge.
    C. Combination square.
    D. Guide for a saw.

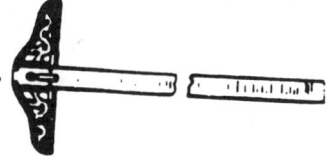

    ANSWER = A

3. The object shown below is a:

    A. Transit knob.
    B. Center punch.
    C. Clock pendulum.
    D. Plumb bob.

    ANSWER = D

4. The object shown below is a:

    A. Drawer pull.
    B. Cleat.
    C. Door stop.
    D. Window lock.

    ANSWER = B

5. The object shown below is a:

    A. Manifold gasket.
    B. Brake band.
    C. Clutch plate.
    D. Air filter.

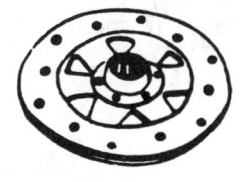

    ANSWER = C

6. The object shown below is a:

    A. paper clip.
    B. wire staple.
    C. Cotter pin.
    D. finishing pin.

    ANSWER = C

7. The object shown below is used to:

   A. Scrape paint.
   B. Etch leather.
   C. Install widow glass.
   D. Drive screws in close areas.

   ANSWER = D

8. The object shown below is used:

   A. For welding.
   B. For soldering.
   C. For lubricating.
   D. For painting.

   ANSWER = A

9. The object shown below is used to:

   A. Measure outside dimensions.
   B. Measure inside dimensions.
   C. Draw circles.
   D. Mark length.

   ANSWER = B

10. The object shown below is used to:

    A. Punch holes.
    B. Clamp objects.
    C. Measure thickness.
    D. Measure depth.

    ANSWER = C

11. The object shown below is used to:

    A. Measure depth.
    B. Measure thickness.
    C. Punch holes.
    D. Test metal hardness.

    ANSWER = A

12. The object shown below is used to:

    A. Set spark plugs.
    B. measure saw blades.
    C. Measure drill bits.
    D. Measure wire.

    ANSWER = D

13. The object shown below is used to:

    A. Refill batteries.
    B. Extinguish fires.
    C. Lubricate vehicles.
    D. Kill insects.

    ANSWER = D

14. The object shown below is used to:

    A. Countersink.
    B. Back screws out.
    C. Remove rivets.
    D. Drill in glass.

    ANSWER = A

15. The object shown below is used to:

    A. Tap pipes.
    B. Turn nuts.
    C. Turn phillip screws.
    D. Round corners of nuts.

    ANSWER = B

16. The object shown below is used to:

    A. Start engines.
    B. Replace spark plugs.
    C. Drill holes.
    D. Change auto tires.

    ANSWER = D

17. The object shown below is used to:

    A. Lift automobiles.
    B. Open oil drums.
    C. Straighten bumpers.
    D. Measure heights.

    ANSWER = A

18. The object shown below is used to:

    A. Thread pipes.
    B. Flare metal tubing.
    C. Pull gears off shafts.
    D. Measure diameters.

    ANSWER = C

SECTION 6

SCIENCE KNOWLEDGE : CHEMISTRY / PHYSICS

EXPLANATION

Entrance level Firefighters exams will usually have some questions relating to Science. The Science questions will usually cover a broad range of subjects pertaining to general science and Fire Science/Chemistry/Physics.

When taking an examination and you encounter science, Chemistry, or physics type questions: don't be afraid if your knowledge is not very broad! General knowledge of basic concepts will be adequate.

If you have knowledge of any foreign languages, use this knowledge when recalling the meanings of technical terms. For example, Latin roots of words are particularly revealing.

Use your common sense when dealing with unfamiliar areas. What seems to be the most sensible answer will normally be the correct answer.

SCIENCE/CHEMISTRY
INFORMATION RELATING TO THE FIRE SERVICE:

B.T.U. = British thermal unit.

B.T.U. = 1 LB water increases 1 degree F.

B.T.U. is the amount of heat required to raise one pound of water one degree F. (at atmospheric pressure).

To convert one pound of ice at 32 degrees F to steam at 212 degrees F requires **1293.7 B.T.U.'s**.

To convert ice to water requires **143.4 B.T.U.'s**.

To convert water to steam requires **970.3 B.T.U.'s**.

To raise 32 degrees F to 212 degrees F requires **180 B.T.U.'s**.

One pound of construction wood = **8120 B.T.U.'s**.

One gallon of water will absorb about **8000 B.T.U.'s**.

SPECIFIC HEAT = number of B.T.U.'s to raise 1 LB substance 1 degree F.

SPECIFIC HEAT = absorption of heat.

LATENT HEAT = absorbed heat or heat given off.

LATENT HEAT = heat absorbed or given off from a substance as it transfers from a liquid to a gas or a solid to a liquid.

SUBLIME: is when a solid changes to vapor without passing through the liquid phase.

VAPOR = substance in a gaseous state, particularly that of liquid or solid at normal temperatures.

When converting liquid to vapor, the **VOLUME IS INCREASED BY 1700 TIMES**.

A gallon of water may produce a maximum of **200 CUBIC FEET OF STEAM**.

VAPOR DENSITY = ratio of gases.

VAPOR DENSITY = the weight of a determined volume of a gas or vapor, as related to the weight of the same volume of normal dry air.

VAPOR DENSITY of air is = 1, gases of less than 1 indicates the gas is lighter than air.

The **RATE OF DIFFUSION** of an unconfined gas varies inversely with its **VAPOR DENSITY**.

SPECIFIC GRAVITY = ratio of weight to volume.

SPECIFIC GRAVITY = the comparison of the weight or mass of a given volume of a substance at a specific temperature to that of an equal volume of another substance.

SPECIFIC GRAVITY applies to liquids only.

SPECIFIC GRAVITY is the ratio of solid weight or liquid to volume of water. Water = 1 at 4 degrees C or 39 degrees F.

MAXIMUM DENSITY of water is at **39 DEGREES F**.

Liquids of a specific gravity of less than 1 will **FLOAT ON WATER**.

Liquids of specific gravity of more than 1, **WATER WILL FLOAT ON TOP** of this liquid.

LIQUIDS will not expand indefinitely like gases.

A **SUBSTANCE** = gas if it is in a gaseous state at 100 degrees F and 40 PSI.

A **SUBSTANCE** = liquid if it is in a liquid state at 70 degrees F and 14.7 PSI.

BOILING POINT = temperature when vapor is equal to atmospheric pressure.

BOILING POINT = the temperature that a liquid will swiftly convert to vapor.

BOILING POINT = temperature of liquid at which vapor is equal to atmospheric pressure.

Most **USEFUL** information concerning hazard of a liquid is the **FLASH POINT**.

FLASH POINT more than any other physical property determines the hazard of a liquid.

FLASH POINT = less than 5 degrees F below fire point.

FLASH POINT = the lowest temperature of liquid at which it gives off vapor which forms ignitible mixture with air at the surface of the liquid or within the vessel it is used. (capable of propagation of flame with heat, fire does not continue with a heat source).

FLASH POINT = ignitable temperature of a liquids vapor.

FIRE POINT = temperature that a **LIQUID** is able to continue to burn without an outside heat source.

FIRE POINT is the lowest temperature of a **LIQUID** at which vapors are evolved fast enough to continue combustion.

IGNITION TEMPERATURE = temperature that a **SUBSTANCE** is able to continue to burn without an outside heat source.

IGNITION TEMPERATURE is the temperature at which a **SOLID COMBUSTIBLE SUBSTANCE** will ignite and burn.

IGNITION TEMPERATURE of combustible natural gas-air mixture is 1000 degrees F.

FLAMMABLE DENSITY = the range of combustible vapors or gas mixtures with air between the upper and lower flammable limits.

FLAMMABLE LIQUIDS are liquids with flash points **BELOW 100 DEGREES F**.

COMBUSTIBLE LIQUIDS are liquids with flash points **AT 100 DEGREES OR HIGHER**.

Combustible liquids are **SAFER** than flammable liquids.

VAPOR PRESSURE OF A LIQUID: is the pressure of the vapor at any given temperature at which the vapor and liquid phases of the substance are in balance in a closed vessel.

Generally as the **BOILING POINT** of a liquid goes down, the **VAPOR PRESSURE** and evaporation rat **INCREASE**.

FOR THE FOLLOWING TWO CHARTS :
   FP = FLASH POINT AND BP = BOILING POINT.

CLASSIFICATION OF FLAMMABLE LIQUIDS:
   Class I   have FP less   than   73 Degrees F.
   Class IA have FP less   than   73 degrees F.
             BP less   than   100 degrees F
   Class IB have FP less   than   73 degrees F.
             BP greater than   100 degrees F.
   Class IC have FP between   73 and 99 Degrees F.

CLASSIFICATION OF COMBUSTIBLE LIQUIDS:
   Class II   have FP between 100 and 140 degrees F.
   Class III   have FP above   140 degrees F.
   Class IIIA have FP between 140 and 199 degrees F.
   Class IIIB have FP above   200 degrees F.

VAPORIZATION: is the process that a substance changes from liquid or solid phase to a gas.

1 degree C = 1.8 degrees F - (+32 degrees F).
Example: 50C = 90F + 32F = 122 degrees F.
THIS IS FOR POSITIVE TEMPERATURES!

1 degree C = 1.8 degrees F - (+32 degrees). Example:
-40C = 72F -(+32F) = 40 degrees F.
THIS IS FOR NEGATIVE TEMPERATURES!

-40 degrees F and -40 degrees C is the only temperature that degrees F and degrees C are **IDENTICAL**.

Steel sparks = **2500 DEGREES F.**

Temperature of a match flame = **2000 DEGREE F.**

Temperature of a cigarette = **550 - 1350 DEGREES F.**

SALT WATER boils at **226 DEGREES F.**

Unprotected steel looses its strength at about **1000 DEGREES F.**

Bare steel has a fire resistance of about **TEN MINUTES**.

Under severe fire conditions, columns of cast iron and unprotected steel collapse in **10 - 20 MINUTES**.

Firefighters should be aware that most metals will **EXPAND WHEN HEATED**.

Wood burns at a rate of **1" IN 45 MINUTES**.

Wood chars or burns at a rate of **ONE INCH IN 45 MINUTES** at an average combustion temperature of 1400 degrees F.

The maximum temperature that wood should be exposed to constantly is **300 DEGREES F.**

WINDOW GLASS will usually melt at **850 DEGREES F.**

Cheapest construction material is **WINDOW GLASS**.

LIGHT BULBS will begin to swell and lose their shape at about **900 DEGREE F.**

Fires usually burn **UPWARD AND OUTWARD**.

In most structure fires, the floor temperatures are usually about **ONE THIRD** that of the ceiling temperatures.

SPONTANEOUS is slow oxidation.

CHEMICAL REACTIONS double their rate with each **18%** rise in temperature.

In this country the **MAJORITY** of fires occur in dwellings confined to **ONE ROOM**.

The leading cause of fire is **SMOKING**.

Dwelling fires usually involve the **KITCHEN**.

ROUGH DARK SURFACES are the best surfaces for absorbing and radiating heat.

Axe handles are usually made of wood rather that metal primarily because the wooden handles will cushion the impact to the Firefighter.

Ice on sidewalks will melt from the application of salt by the lowering of the freezing point of the water by the salt.

The main hazard of static electricity is the possibility of an explosion created by sparks.

In warm climate areas, water temperature may be lowered by storing the water in earthenware containers, because some of the water changes to vapor, which lowers the water temperature.

Usually the color and odor of smoke will indicate the kind of material that is burning.

A wet hand will stick to a piece of metal but not to a piece of wood, at certain temperatures, because metal is a better conductor of temperature.

The height of waves may be reduced by pouring oil on them, because the surface tension will be weakened.

Substances that are subject to spontaneous combustion are capable of igniting without an external source of heat.

The air that humans breath out has more water vapor in it than the air that humans breath in.

If the formation of ice takes place within a closed container, the container will burst because the water expands when it freezes, which builds up great pressure.

Kerosene lamps burn with a yellow flame because of the incandescence of unburned carbon particles.

When metal pipe is used to carry liquids or gases, the threaded portion is always tapered.

Smoke usually rises from a fire because the cooler, heavier air displaces the lighter warm air.

Water heated at sea level will boil at a higher temperature than water heated at high altitudes.

Good conductors of heat are usually poor insulators of heat.

Chemically pure water may be made from tap water by the process of distillation.

When air rises it cools because it expands.

A catalyst is a substance that will change the speed of a chemical reaction.

The sprocket wheels and chain of a bicycle increase the speed of the rear wheel. The height at which an object will float on water is determined mainly by the weight of the water displaced by the object.

The exhaust gas of an internal combustion engine is mainly carbon, carbon dioxide, carbon monoxide, nitrogen, and steam.

SPECIFIC GRAVITY = the weight or mass of a given volume of a substance at a specified temperature, as compared to that of an equal volume of another substance.

Flammable liquid vapors have **VAPOR DENSITY** of greater than 1, air = 1 therefore they will sink in the atmosphere.

VAPOR DENSITY = the weight of a vapor-air mixture resulting from the vaporization of a flammable liquid at equilibrium temperature and pressure conditions, as compared with the weight of an equal volume of air under the same conditions.

FLASH POINT more than any other property determines the hazard of the liquid.

FLASH POINT is less than **5 DEGREES BELOW** fire point.

FLASH POINT is the minimum temperature at which a liquid gives off sufficient vapor to form an ignitible mixture with air.

FIRE HYDRAULICS

Four **FUNDAMENTALS** that govern friction loss in hose lines and pipes:
1. All other conditions being equal the loss by friction varies with the length of the line.
2. In the same size hose, friction loss varies directly as the square of velocity flow.
3. For the same flow, friction loss varies inversely as the fifth power of the diameter of the hose.
4. For a given velocity flow, friction loss is independent of the pressure.

Friction loss in two lines of 2 1/2" hose is about **28%** of the friction loss in a single line of 2 1/2" hose.

Friction loss in old hose may be **50%** greater than new hose.

Friction loss is governed by the **QUANTITY (GPM)** of water flowing.

Hose in a **ZIG-ZAG** pattern will increase friction loss by about **6%**.

If the length of hose is **DOUBLED** the friction loss is **DOUBLED**.

Pressure **DOES NOT** effect friction loss in the same size hose. (increase or decreases)

MERCURY weighs **.49 LBS** per cubic inch.
  EXAMPLE:
    30 Hg" indicates 30 X .49 =
    14.7 LBS atmospheric pressure.

NORMAL ATMOSPHERIC PRESSURE:
    29.92 Hg" or 14.7 PSI; (29.92 X .49 = 14.7)

PERFECT VACUUM; water could theoretically allow atmospheric pressure to raise water 33.9 feet. (29.92 Hg" X 1.13 = 33.92)

ATMOSPHERIC PRESSURE of 14.7 PSI is capable of sustaining a column of water 33.9 feet high (14.7 X 2.304 = 33.9), this is if a pump could produce a perfect vacuum. (the practical limit is about 22 feet).

ATMOSPHERIC PRESSURE decreases as elevation increases at a rate of approximately 1/2 LB every 1000 feet of elevation.

PRESSURE may be defined as the measurement of energy contained in water.

The **THEORETICAL HEIGHT** to which a pumper can raise water by suction decreases proportionally as elevation increases, therefore the theoretical height to which a pumper can raise water decreases about 1 foot every 1000 feet of altitude.

PRESSURE, as we normally think of it in the fire service is that force delivered by pumpers for supplying water to a fire for firefighting.

BACK PRESSURE = .434 X height. (for field use .5 PSI) or 1/2 of the height = head.

PRESSURE: force per unit area measured in PSI.

BACK PRESSURE: the pressure from a static head of water.

Normally **PRESSURE** works for us, but when pumping up hills or to nozzles located on upper floors of buildings, it can work against us. (back pressure).

HEAD is the vertical distance from the surface of the water being considered to the point being considered.

If the nozzle is to be lower than the pump, the factor of elevation must be **SUBTRACTED** from the pump pressure.

NOZZLE PRESSURE on a charged line with nozzle closed is approximately equal to the same pressure as the engine pressure. (+ or - the head)

NOZZLE PRESSURE: the velocity pressure in PSI at which water is discharged from a nozzle.

NOZZLE REACTION is caused by an increase in water velocity.

VELOCITY FLOW is the speed at which water passes a given point.

If tip size is **DOUBLED** the nozzle reaction will be **FOUR** times as great.

Friction loss is only affected by water if the pressure **INCREASES THE VELOCITY** of the water.

Pressure has **NO EFFECT** on the friction loss as long as it does not change the velocity of the flow. It is the **VELOCITY** of the water over the lining of the hose and the condition of the lining that governs the friction loss.

Basically without turbulence and other factors, the velocity of a water stream when it leaves the outlet of a nozzle depends on the **PRESSURE AT THE OUTLET ONLY.**

If the nozzle on a long line of hose is replaced by a smaller nozzle and the engine pressure remains the same, the nozzle pressure will be **INCREASED**.

WATER weighs approximately **62.5** LBS per cubic foot.

One cubic foot of **WATER** contains **1728** cubic inches.

One gallon of **WATER** contains **231** cubic inches.

One gallon of **WATER** weighs **8.35** LBS.

There are **7.481** gallons in a cubic foot.

A column of **WATER** **1"** high and **1"** square weighs **.434** LBS. (note .5 PSI for field calculations).

A column of **WATER** **2.304** feet in height, will exert **1** LB PSI at its base.

HEAD is the vertical distance from the surface of the water being considered to the point being considered.

PHYSICS

PHYSICS is the science that deals with energy and matter.

DYNAMICS is the general term used to describe the laws that govern forces in which motion is produced.

CENTRIFUGAL FORCE is the tendency of an object traveling in a curve to try and go in a straight line.

GRAVITATION is the physical occurrence of weight that an object maintains.

GRAVITY is the term used to describe the attraction that exist between the earth and the bodies on or near it.

WEIGHT is the measure of the earths attraction for a mass.

The loss of weight of an object submerged in a liquid is equal to the weight of the displaced liquid.

One cubic foot of water weighs 62.4 pounds.

Scissors are an example of a first class **LEVER**.

It is easier to turn a wrench with a long handle than a short handle, because it provides more **LEVERAGE**.

FULCRUM is the fixed point on which a lever turns when it is used.

The **SCREW** is a powerful machine capable of moving large objects.

SPECIFIC GRAVITY is the weight of a substance compared to an equal amount of water.

The most common way to determine **SPECIFIC GRAVITY** is with the use of a **HYDROMETER**.

FRICTION is the force which tends to keep an object, placed on an incline, from slipping down.

FRICTION is the resistance encountered when an object moves over another object.

FLUID FRICTION depends upon the speed of the flow.

The type of electricity produced by **FRICTION** is called **STATIC ELECTRICITY**.

Machinery is oiled in order to decrease **FRICTION** of moving parts.

For every **ACTION** there is a **REACTION**.

NEWTON discovered the law of inertia.

HOOKE'S LAW states that strain is proportional to stress.

RADIUS is the distance from the center to the edge of a circle or a sphere.

CIRCUMFERENCE is the line around a circle.

ANGLE is the space between two lines or surfaces that meet at a point or that cross each other.

ACUTE ANGLE is an angle that is smaller than a right angle, less than 90 degrees.

RIGHT ANGLE is an angle which is at 90 degrees.

OBTUSE ANGLE is an angle that is greater than a right angle, more than 90 degrees.

ANGLES are measured in terms of degrees.

SHEAR is a type of stress.

When a **MAGNET** is set into a mound of iron shavings, the shavings will cling near the ends.

The most common **MAGNETIC** substance is **IRON**.

ARMATURE is an iron core that will rotate within pole pieces.

The principle parts of an electric motor are a stationary field **ELECTROMAGNET** and an **ARMATURE MAGNET**.

A permanent **MAGNET** is used in a speedometer.

DYNAMO is a machine that converts electrical energy into mechanical energy.

Silver is one of the metal substances that **MAGNETIZATION** is least easily produced.

TURBINE is a type of engine operated by the pressure of water, steam, or air on ridges or fins, called vanes, on the side of a rotating disk.

The principle for the operation of a **STEAM ENGINE** is based on steam compression.

The capacity of a **STEAM BOILER** is the amount of steam it can produce in an **HOUR**.

BIMETALLIC THERMOSTAT function because of the fact that different metals expand at different rates when heated.

CONDUCTOR is any material which will allow an electric current to pass through it.

TRANSFORMER will change the voltage of an alternating electrical current.

ALTERNATING CURRENT is the electrical current that reverses its polarity at rapid regular intervals.

INSULATOR is a substance that offers high resistance to the passage of electricity.

CONDENSER is able to stop direct currents.

PORCELAIN is a very good electrical insulator.

AMPERES is the term used to describe the strength of electrical current.

PRESSURE is the instrument which causes electrical current to flow through wire.

BAROMETER is an instrument that measures atmospheric pressure.

When water freezes it **EXPANDS**.

SUCTION is created by **VACUUM**.

Liquid will rise into a straw, into your mouth, and upwards because the pressure on the liquid is greater than the pressure in the mouth.

BAROMETER is the instrument that is used to determine the pressure of air changes.

ELECTRON is a particle of negative electricity.

PROTONS are the positive particles of electricity contained within atoms.

ELECTROLYSIS is the process of decomposing water by electricity.

HEAT will make the molecules in ice move faster.

FUEL is whatever is burned to create **HEAT**.

CALORIE is the unit of **HEAT**.

HEAT and **LIGHT** pass more than ninety millions of miles from the sun to the earth by radiation.

LIGHT travels from the sun to the earth in about eight seconds.

HEAT passes through iron by conduction.

Metal placed into a hot liquid will **EXPAND** because of conduction.

ELEMENTS contain heat energy, and produce heat while they are uniting.

SOUND travels at about 1090 feet per second.

GYROSCOPE is a device that is used to increase stability.

FOOT POUND is the unit applied to **WORK**.

ADIABATIC is a change in the condition of a material that occurs without gain or loss of heat from surrounding material.

KINETIC PRESSURE is moving pressure or pressure in motion, such as water moving through hose lines under pressure.

SCIENCE/CHEMISTRY/PHYSICS QUESTIONS

REMEMBER: read the question and all answer choices carefully prior to determining the correct answer. Go on to the next question if you cannot select an answer quickly. Actual exams are timed.

1. Which of the following statements is correct:
   A. Water heated at sea level will boil at a higher temperature than water heated on the top of a mountain.
   B. A large quantity of water will boil at a higher temperature than a small quantity.
   C. Water heated slowly by a low flame will boil at a higher temperature than water heated quickly by a high flame.
   D. Water always boils at the same temperature regardless of pressure.

   ANSWER = A

2. Earthenware jugs will keep water cool, even in hot temperatures, because:
   A. Particles of earthenware dissolve in the water and lower the waters temperature.
   B. The rough surface of the jug radiates heat more rapidly.
   C. The change of some of the water to a vapor lowers the temperature.
   D. The jug is a good conductor of heat.

   ANSWER = C

3. Of the following toxic gases, the one which is the most dangerous because it cannot be seen and has no odor is:
   A. Ammonia.
   B. Chlorine.
   C. Carbon monoxide.
   D. Ether.

   ANSWER = C

4. If oil poured on water it will reduce the height of the waves, because:
   A. Oil fills up the toughs of the waves.
   B. Chemical action between oil and water produces a heavier substance.
   C. Added weight of the oil makes the waves break at a lesser height.
   D. Surface tension of the water will be weakened.

   ANSWER = D

5. When kerosene burns with a yellow flame, this is due to the:
   A. Burning of hydrogen.
   B. Incandescence of unburned carbon particles.
   C. Complete burning of the hydrocarbons.
   D. Heating of the wick.

   ANSWER = B

6. The threads on metal pipe that is used for liquids and gases will always be:
   A. Straight.
   B. Fine.
   C. Coarse.
   D. Tapered.

   ANSWER= D

7. The following principle that applies to a jet plane is:
   A. Energy can be neither created or destroyed.
   B. If pressure is applied to a confined gas, the volume of the gas will decrease.
   C. For every action there is an equal and opposite reaction.
   D. Every effect has a cause.

   ANSWER = C

8. Substances that submit a great deal of resistance to the passage of electricity through it is called:
   A. Fuse.
   B. Insulator.
   C. Conductor.
   D. Transformer.

   ANSWER = B

9. Why is it beneficial to enter a smoke filled room crawling low to the floor?
   A. Because smoke will be radiated through the floor.
   B. Because smoke is lighter than air.
   C. Because smoke may become flame.
   D. Because smoke will be compressed in a room.

   ANSWER = B

10. A tank that weighs 500 pounds in the air appears to weigh 300 pounds when at a depth of 10 feet under water. If the tank is lowered to a depth of 20 feet, what will its apparent weight be?
    A. 100 pounds.
    B. 200 pounds.
    C. 300 pounds.
    D. 400 pounds.

    ANSWER = C

11. Of the following which is least likely to explode:
    A. Shut can of water over a high heat source.
    B. A can full of gasoline.
    C. A can of gasoline that is half empty.
    D. Gasoline can that has just been emptied.

    ANSWER = B

12. The main reason to insulate pipes in heating:
    A. Is to prevent the loss of heat.
    B. Is to prevent pipes from rusting.
    C. Is to prevent fires.
    D. Is to prevent injury to people.

    ANSWER = A

13. Which of the following is the best heat conductor:
    A. Copper.
    B. Glass.
    C. Steel.
    D. Water.

    ANSWER = A

14. Which of the following represents the freezing point on a Fahrenheit thermometer:
    A. Zero degrees.
    B. 12 degrees.
    C. 30 degrees.
    D. 32 degrees.

    ANSWER = D

15. An electron is:
    A. An electronic nucleus.
    B. A hydrogen atom.
    C. A static charge.
    D. A particle of negative electricity.

    ANSWER = D

16. The exhaust gasses of an automobile consist of which of the following gases:
    A. Steam.
    B. Carbon, carbon dioxide, and carbon monoxide.
    C. Nitrogen.
    D. All of the above.

    ANSWER = D

17. Electrical pressure is expressed in terms of:
    A. Ampere.
    B. Pressure gauge.
    C. Volt.
    D. Ohm.

    ANSWER = C

18. The chain and the sprocket wheels of a bicycle increase the:
   A. Power of the rider.
   B. Force applied to the rear wheel.
   C. Force applied to the road.
   D. Speed of the rear wheel.

   ANSWER = D

19. Of the following, what happens to water when it freezes:
   A. It contracts.
   B. It changes chemically.
   C. It expands.
   D. It increases substantially.

   ANSWER = C

20. What is the term used to define the attraction that exist between the earth and the bodies on or near it:
   A. Force.
   B. Leverage.
   C. Vibration.
   D. Gravity.

   ANSWER = D

21. Water will conduct electrical current better when:
   A. Its mineral content is at a high level.
   B. Its organic content is at a high level.
   C. Its mineral content is at a low level.
   D. It is completely pure.

   ANSWER = A

22. Which of the following represents the weight of water:
   A. 12.50 pounds.
   B. 8.33 pounds.
   C. 5.50 pounds.
   D. 4.33 pounds.

   ANSWER = B

23. Assume that you have two containers of equal size and shape. They are filled with equal amounts of water. A block of ice is added to one container and the same quantity of ice, chopped into cubes is added to the other container. What will happen to the water in the container with the cubes as compared to the container with the block?
   A. It will cool to the same temperature, faster.
   B. It will cool to the same temperature, slower.
   C. It will cool to a lower temperature, faster.
   D. It will cool to a lower temperature, slower.

   ANSWER = A

24. Of the following, which is the worst substance to have contact a rope used for firefighting uses?
   A. Water.
   B. Oil.
   C. Gasoline.
   D. Acid.

   ANSWER = D

25. Which science does hydraulics pertain to?
   A. Water/liquids under pressure.
   B. Water/liquids in motion.
   C. Gravity of water.
   D. Water for firefighting.

   ANSWER = B

26. What is the substance that is usually used for electrical conductors?
   A. Rubber.
   B. Glass.
   C. Porcelain.
   D. Copper.

   ANSWER = D

27. What is the term used to express the power of attraction of one chemical for another?
   A. Affinity.
   B. Oxidation.
   C. Gravity.
   D. Ventilation.

   ANSWER = A

28. Materials that are incapable of combustion are:
   A. Flammable.
   B. Fireproof.
   C. Fire hard.
   D. Incinerated.

   ANSWER = B

29. What would be the most likely cause of ignition to a pile of rags and trash laying in the corner of a room?
   A. Arson.
   B. Arching wires.
   C. Cigarette butt.
   D. Spontaneous combustion.

   ANSWER = D

30. Petroleum is:
   A. Used only as a lubricant.
   B. Made from gasoline.
   C. Lighter than water.
   D. Used only as fuel.
   ANSWER = C

31. Oils value as a lubricant depends upon its:
   A. Mixture.
   B. Origin.
   C. Viscosity.
   D. Color.

   ANSWER = C

32. Which of the following would represent the greatest threat to an explosion:
   A. Square tanks.
   B. Round tanks.
   C. Full tanks.
   D. Almost empty tanks.

   ANSWER = D

33. What unit of measurement does a millimeter represent?
   A. Volume.
   B. Weight.
   C. Viscosity.
   D. Length.

   ANSWER = D

34. One millimeter equals what portion of a meter?
   A. One-thousandth of a meter.
   B. One-tenth of a meter.
   C. One half of a meter.
   D. One meter.

35. What does the electrical term "arching" refer to?
   A. An electrical current flowing along a conductor bent into a curve.
   B. An electrical current passing across a gap between two conductors.
   C. The blowing out of an electrical fuse.
   D. The blowing out of a circuit breaker.

   ANSWER = B

36. When a material is referred to as combustible, it most likely is:
   A. Fireproof.
   B. Flammable.
   C. Crammed.
   D. Disintegrated.

   ANSWER = B

37. Fire hose should not be cleaned with the use of gasoline because:
   A. It would cause it to rot.
   B. It would weaken the couplings.
   C. It would stain it.
   D. It would shrink it.

   ANSWER = A

38. The smallest unit that a substance may be divided into without destroying the substance is:
   A. Atom.
   B. Particle.
   C. Nucleus.
   D. Molecule.

   ANSWER = D

39. Which of the following oil soaked rags is most susceptible to spontaneous ignition:
   A. Motor oil.
   B. Castor oil.
   C. Linseed oil.
   D. Fuel oil.

   ANSWER = C

40. The occurrence of unusual percentages of carbon monoxide at smoldering fires would most likely be due to:
   A. Oxygen deficiency.
   B. Very low temperature.
   C. Very high temperature.
   D. Similarity of materials.

   ANSWER = A

41. When oxygen concentration in the air drops below 16 per cent:
   A. No flames will continue to burn.
   B. Most flames will be extinguished.
   C. Most materials will continue to burn freely.
   D. Breathing will not be affected.

   ANSWER = B

42. When radioactive materials are involved in fires, the radiation coming from the fire:
   A. Increases in rate and intensity.
   B. Decreases in rate and intensity.
   C. Increases in rate and decreases in intensity.
   D. Is not affected.

   ANSWER = D

43. As a fire increases in temperature, the percentage of carbon monoxide in the gases liberated:
   A. Remains constant.
   B. Increases.
   C. Decreases.
   D. Decreases at a fast rate.

   ANSWER = B

44. Geiger counters will react to:
    A. Both alpha and beta, but not gamma radiation.
    B. Both alpha and gamma, but not beta radiation.
    C. Both beta and gamma, but not alpha radiation.
    D. Alpha, beta, and gamma radiation.

    ANSWER = C

45. Hot magnesium, reacting with water, liberates:
    A. Manganous chloride.
    B. Nitrous oxide.
    C. Hydrogen.
    D. Nitrogen.

    ANSWER = A

46. Natural gas consist principally of:
    A. Propane.
    B. methane.
    C. ethane.
    D. Butane.

    ANSWER = B

47. Of the following flame colors, which one would indicate the highest temperature:
    A. Orange-red.
    B. Light red.
    C. orange-yellow.
    D. Yellow-white.

    ANSWER = D

48. The freezing point on a Fahrenheit thermometer is:
    A. Zero degrees.
    B. 10 degrees.
    C. 28 degrees.
    D. 32 degrees.

    ANSWER = D

49. An electron is a/an:
    A. Particle of negative electricity.
    B. Static charge.
    C. Hydrogen atom.
    D. Electronic nucleus.

    ANSWER = A

50. Any substance which offers a very high or very great resistance to the passage of electricity through it is called a/an:
    A. Fuse.
    B. Conductor.
    C. Insulator.
    D. Closed circuit.

    ANSWER = C

SECTION 7

MATH CONCEPTS / ARITHMETIC

EXPLANATION

The Math concepts and arithmetic areas that are usually covered on Entrance level Firefighter exams include:
1. Addition.
2. Subtraction.
3. Multiplication.
4. Division.
5. Percentages.
6. Decimals.
7. Fractions.
8. Word Problems.

When encountering the math portion of the exam: work fast but don't concede precision for swiftness. Don't jump to conclusions when computing the answers. Remember when you work out problems in your head it is easy to make a mistake.

Remain calm, even if the problem looks complicated at first. You have seen these types of problems before. Don't panic it will come back to you. If you cannot answer a question, go on to the next one and return to the difficult ones later if you have the time.

ADDITION PROBLEMS

ADD:

1. 42 + 33 + 18 =

    A. 83   B. 93   C. 91   D. 79   E. None

    ANSWER = B

2. 18 + 16 + 9 =

    A. 39   B. 43   C. 49   D. 53   E. None

    ANSWER = B

3. 53 + 26 + 17 =

    A. 86   B. 89   C. 93   D. 96   E. None

    ANSWER = D

4. 5 + 13 + 88 + 102 =

    A. 198   B. 208   C. 193   D. 203   E. None

    ANSWER = B

5. 58 + 62 + 104 + 153 =

    A. 348   B. 357   C. 376   D. 377   E. None

    ANSWER = D

6. .55 + .34 =

    A. .86   B. .87   C. .89   D. .90   E. None

    ANSWER = C

7. .96 + .88 =

    A. 1.88   B. 1.86   C. 1.84   D. 1.80   E. None

    ANSWER = C

8. 1.10 + .99 =

   A. 1.99   B. 2.09   C. 2.19   D. 2.28   E. None

   ANSWER = B

9. 21.9 + 2.3 =

   A. 23.3   B. 23.9   C. 24.2   D. 24.4   E. None

   ANSWER = C

10. 88.8 + 99.9 =

    A. 178.8   B. 187.8   C. 188.7   D. 188   E. None

    ANSWER = C

11. 3/4 + 3/4 =

    A. 1   B. 1 1/4   C. 1 1/2   D. 1 1/34   E. None

    ANSWER = C

12. 1/2 + 1/4 =

    A. 3/4   B. 1   C. 1 1/4   D. 1 1/2   E. None

    ANSWER = A

13. 1/3 + 1 1/3 =

    A. 2/3   B. 1   C. 1 1/2   D. 1 2/3   E. None

    ANSWER = D

14. 5/8 + 7/8 =

    A. 1   B. 1 1/8   C. 1 1/2   D. 1 5/8   E. None

    ANSWER = C

15. 1/5 + 1/10 =

    A. 2/5   B. 3/5   C. 4/5   D. 1   E. None

    ANSWER = B

SUBTRACTION PROBLEMS

SUBTRACT:

1. 93 − 64 =
   A. 28   B. 29   C. 31   D. 38   E. None

   ANSWER = B

2. 48 − 16 =
   A. 28   B. 29   C. 32   D. 33   E. None

   ANSWER = C

3. 77 − 65 =
   A. 12   B. 14   C. 15   D. 16   E. none

   ANSWER = A

4. 98 − 29 =
   A. 66   B. 68   C. 73   D. 74   E. None

   ANSWER = E

5. 142 − 133 =
   A. 7   B. 8   C. 9   D. 10   E. None

   ANSWER = C

6. .55 − .46 =
   A. .06   B. .07   C. .08   D. .09   E. None

   ANSWER = D

7. .99 − .79 =
   A. 20   B. 2.0   C. .22   D. .20   E. None

   ANSWER = D

8. 1.08 − .76 =
    A. .23   B. .32   C. .33   D. .34   E. None

    ANSWER = B

9. 22.64 − 22.46 =
    A. .18   B. 1.8   C. 18   D. 1.3   E. None

    ANSWER = A

10. 132.1 − 113.9 =
    A. 18   B. 18.2   C. 12.8   D. 12   E. None

    ANSWER = B

11. 3/4 − 3/4 =
    A. 1/4   B. 1/3   C. 1/8   D 1/10   E. None

    ANSWER = E

12. 3/4 − 1/4 =
    A. 1/4   B. 1/3   C. 1/2   D. 3/4   E. None

    ANSWER = C

13. 1/3 − 1/4 =
    A. 1/8   B. 1/6   C. 1/10   D. 1/12   E. None
    ANSWER = D

14. 1 1/4 − 1 1/8 =
    A. 1/4   B. 1/8   C. 1/3   D 1/6   E. None
    ANSWER = B

15. 22 1/2 − 15 1/6 =
    A. 7   B. 7 1/6   C. 7 1/2   D. 7 1/8
    ANSWER = D

MULTIPLICATION PROBLEMS

MULTIPLY:

1. 3 X 87 =
   A. 251   B. 255   C. 261   D. 266   E. None

   ANSWER = C

2. 6 X 58 =
   A. 348   B. 358   C. 361   D. 364   E. None

   ANSWER = A

3. 9 X 98 =
   A. 872   B. 874   C. 878   D. 882   E. None

   ANSWER = D

4. 21 X 44 =
   A. 904   B. 914   C. 924   D. 925   E. None

   ANSWER = C

5. 18 X 111 =
   A. 1998   B. 1999   C. 2008   D. 2011   E. None

   ANSWER = A

6. .25 x .25 =
   A. .0625   B. .625   C .602   D. .620   E. None

   ANSWER = A

7. .5 X .5 =
   A. .10   B. .20   C. .25   D. .30   E. None

   ANSWER = C

8. .325 x .20 =

   A. 6.5   B. .65   C. .065   D. .0065   E. None

   ANSWER = C

9. 1.5 X 1.5 =

   A. 2.05   B. 2.25   C. 2.50   D. 2.55   E. None

   ANSWER = B

10. .1 X .1 =

   A. 1   B. .1   C. .01   D. .11   E. None

   ANSWER = C

11. 1/4 X 1/4 =

   A. 1/8   B. 1/16   C. 1/24   D. 1/32   E. None

   ANSWER = B

12. 1/8 X 1/4 =

   A. 1/8   B. 1/16   C. 1/24   D. 1/32   E. None

   ANSWER = D

13. 3/4 X 3/4 =

   A. 1/2   B. 9/32   C. 9/16   D. 3/4   E. None

   ANSWER = C

14. 7/8 X 1/4 =

   A. 7/8   B. 7/16   C. 7/32   D. 7/64   E. None

   ANSWER = C

15. 5/8 X 4/5 =

   A. 3/8   B. 1/2   C. 5/8   D. 3/4   E. None

   ANSWER = B

DIVISION PROBLEMS

DIVIDE:

1. 64 by 8 =
   A. 7   B. 8   C. 9   D. 10   E. None

   ANSWER = B

2. 55 by 5 =
   A. 8   B. 9   C. 10   D. 11   E. None

   ANSWER = D

3. 48 by 6 =
   A. 6   B. 8   C. 9   D. 10   E. None

   ANSWER = B

4. 42 by 7 =
   A. 4   B. 5   C. 6   D. 8   E. None

   ANSWER = C

5. 108 by 12 =
   A. 9   B. 12   C. 15   D. 17   E. None

   ANSWER = A

6. 6.4 by 3.2 =
   A. 1.5   B. 1.6   C. 1.8   D. 2   E. None

   ANSWER = D

7. 3.3 by 1 =
   A. 1.3   B. 3.1   C. 3   D. 3.3   E. None

   ANSWER = D

8. 14.44 by 7.22 =
   A. 7.22 B. 7 C. 3.22 D. 3 E. None

   ANSWER = E

9. .444 by 2.2 =
   A. 2.02 B. 20.2 C. .202 D. .020 E. None

   ANSWER = C

10. .15 by .1 =
    A. 15 B. 1.5 C. .15 D. .015 E. None

    ANSWER = B

11. 3/4 by 3/4 =
    A. 1/2 B. 3/4 C. 1 D. 1 1/2 E. None

    ANSWER = C

12. 3/4 by 2/3 =
    A. 1 1/6 B. 1 1/3 C. 1 1/8 D. 1 E. None

    ANSWER = C

13. 1/8 by 1/16 =
    A. 2/16 B. 1/8 C. 1 D. 2 E. None
    ANSWER = D

14. 7/8 by 1/16 =
    A. 11 B. 1.1 C. 14 D. 1.4 E. None
    ANSWER = C

15. 1/2 by 1/3 =
    A. 1 1/2 B. 1 1/3 C. 1 D. 7/8 E. None
    ANSWER = A

PERCENTAGE PROBLEMS

PERCENT =

1. 10% of 30 =
   A. 3   B. 3.3   C. 10   D. 20   E. None

   ANSWER = A

2. 15% of 60 =
   A. 7   B. 8   C. 9   D. 13   E. None

   ANSWER = C

3. 25% of 300 =
   A. 50   B. 55   C. 65   D. 75   E. None

   ANSWER = D

4. 20% of 30 =
   A. 6   B. 15   C. 3   D. 10   E. None

   ANSWER = A

5. 75% of 96 =
   A. 52   B. 57   C. 77   D. 82   E. None

   ANSWER = E

6. 9 is what per cent of 12 ?
   A. 3/4%   B. 7.5%   C. 75%   D. 25%   E. None

   ANSWER = C

7. 10 is what per cent of 30 ?
   A. 2/3%   B. 33%   C. 1/3%   D. 15%   E. None

   ANSWER = B

8. 6 is what per cent of 60 ?
   A. 8% B. 80% C. 10% D. 1.1% E. None
   ANSWER = C

9. 10 is what per cent of 20 ?
   A. 10% B. 20 % C. 45% D. 50 %
   ANSWER = D

10. 90 is what per cent of 200 ?
    A. 90% B. 99% C. 45% D. 49% E. None
    ANSWER = C

11. 100 is what per cent of 300 ?
    A. 33% B. 34% C. 35% D. 36% E. None
    ANSWER = A

12. 99 is what per cent of 100 ?
    A. 1% B. 9% C 90 % D. 99% E. None

    ANSWER = D

13. 46 is what per cent of 200 ?
    A. 23% B. 46% C. 92% D. 50% E. None

    ANSWER = A

14. 50 is what per cent of 200 ?
    A. 50% B. 40% C. 25% D. 15% E. None

    ANSWER = C

15. 80 is what per cent of 400 ?
    A. 60% B. 80% C. 20% D. 40% E. None

    ANSWER = C

WORD PROBLEMS

1. The money spent by Captain Sutton for class room study was as follows: $2.00 for paper; $2.29 for notebook; $1.00 for pencils; 99 cents for a ruler; 29 cents for an eraser and 39 cents for sales tax. What was his total bill?
   A. $6.39
   B. $6.46
   C. $6.69
   D. $6.96

   ANSWER = D

2. Fireman Simpson works in a machine shop on his off duty days. During one month he made the following number of machined parts: 598; 699; 750; 433; 1,200. how many parts did he make?
   A. 3,633.
   B. 3,680.
   C. 3,733.
   D. 3,780.

   ANSWER = B

3. Firefighter Simpson went to the market with a $20 bill. He bought the following items: roast, $7.87; potatoes, $2.34; milk, $1.88; salad fixings $4.65; ice cream, $2.69. How much change did he bring back to the fire station?
   A. 57 cents.
   B. 61 cents.
   C. 66 cents.
   D. 87 cents.

   ANSWER = A

4. The City Fire Departments area is 298 square miles. The County Fire Departments area is 2,342 square miles. How much larger is the County Fire Department area?
   A. 1,942 square miles.
   B. 2,034 square miles.
   C. 2,044 square miles.
   D. 2,054 square miles.

   ANSWER = C

5. The City Fire Department can buy the fire apparatus for $210,354 if it pays cash. If the City Fire Department takes 12 months to pay, it will cost $223,989. How much will they save by paying cash?
   A. $13,365.
   B. $13,635.
   C. $13,656.
   D. $13,665.

   ANSWER = B

6. The year 1991 is the 68th anniversary of the City Fire Department. When was the City Fire Department founded?
   A. 1867.
   B. 1888.
   C. 1893.
   D. 1896.

   ANSWER = C

7. The Firefighter Association's telephone bill listed the following items: $ 87.50 service for long distance; 890 message units at 5 cents each; 22.34 service charge. What was the total bill?
   A. $134.54.
   B. $143.34.
   C. $154.34.
   D. $154.43.

   ANSWER = C

8. Engine 82 is going to the mechanic for re-power. If the trip covers 644 miles round trip and the engine can get 7 miles per gallon at a cost of $1.39 per gallon for fuel. What will be the cost of the fuel for the trip?
   A. $127.88.
   B. $128.87.
   C. $137.88.
   D. $148.87.

   ANSWER = A

9. Engine 88 uses about 6 gallons of fuel for each mile. If it has been fueled with 50 gallons, approximately how far can it go?
   A. 200 miles.
   B. 250 miles.
   C. 300 miles.
   D. 350 miles.

   ANSWER = C

10. A strike team leaves for a brush fire at 2115 hours and arrives at the scene at 2445 hours. The strike team averaged 31 miles per hour. How many miles did they travel?
    A. 105.8 miles.
    B. 108.0 miles.
    C. 108.5 miles.
    D. 118.5 miles.

    ANSWER = C

11. During three 24 hour shift periods: E-81 traveled 66 miles, E-82 traveled 88 miles, E-83 traveled 75 miles, T-81 traveled 33 miles, T-82 traveled 44 miles, R-81 traveled 106 miles, R-82 traveled 155 miles, and C-81 traveled 64 miles. How much more or less than 1,000 miles did there combined total equal?
    A. 369 miles more.
    B. 369 miles less.
    C. 639 miles more.
    D. 639 miles less.

    ANSWER = C

12. Engineer Harper has four lengths of 2 1/2" hose in the following lengths: 49 feet, 50 feet, 48 feet, and 47 feet. What is the total length of these pieces?
    A. 147 feet.
    B. 149 feet.
    C. 194 feet.
    D. 197 feet.

    ANSWER = C

13. During one year the City Fire Departments HQ station had 5,566 total responses. During the same period Station #2 had 6,434 total responses. How many more responses did Station #2 have than HQ?
    A. 686.
    B. 688.
    C. 866
    D. 868.

    ANSWER = D

14. During the first half of the year, the City Fire Department spent 75 per cent of its $1,000,000.00 budget. What is the total amount of monies available from the budget for the remainder of the year?
    A. $250,000.00.
    B. $275,000.00.
    C. $300,000.00.
    D. $325,000.00.

    ANSWER = A

15. Firefighter Jones makes $1,500.00 per pay period. He saves 6 per cent of his earnings each pay period. How much does he save each pay period?
    A. $66.00.
    B. $69.00.
    C. $90.00.
    D. $96.00.

    ANSWER = C

16. The City Fire Department last year required each hazardous business owner to pay $.0088 of their profits for permits to do business within the City. If the profit of a business was a total of $15,000.00, how much did they pay for a permit?
    A. $103.00.
    B. $113.00.
    C. $122.00.
    D. $132.00.

    ANSWER = D

17. B/C Smith paid $.75 per pound for "Spill-Kill". He purchased 12 bags and each bag weighed 25 pounds. What was the total cost?
    A. $200.00.
    B. $205.00.
    C. $210.00.
    D. $225.00.

    ANSWER = D

18. Firefighter Simpson called in sick with a temperature of 101.3 degrees. If the normal temperature is 98.6 degrees, how many degrees above normal was he?
    A. 2.2 degrees.
    B. 2.5 degrees.
    C. 2.7 degrees.
    D. 3.3 degrees.

    ANSWER = C

19. What will be the cost of 300 feet of fire hose at a cost of $1.457 per foot?
    A. $437.10.
    B. $431.70.
    C. $423.10.
    D. $418.70.

    ANSWER = A

20. In three successive shifts, "A" shift spent $34.19, $28.16, and $42.22 for meals. What was the total cost for the meals during this period?
    A. $103.17.   C. $113.07.
    B. $105.17.   D. $117.03

    ANSWER = B

21. Firefighter Dixon bought a pen, paper, and envelopes. The entire bill was $7.50. The pen cost $2.39 and the paper was $3.25. What was the cost of the envelopes?
    A. $1.36.   C. $1.66.
    B. $1.56.   D. $1.86.

    ANSWER = D

22. In a promotional exam, Engineer Harper answered 95% of the questions correctly. There were 140 questions. How many questions did he get right?
A. 139.
B. 133.
C. 127.
D. 123.

ANSWER = B

23. The City Fire Departments baseball team played 20 games last season and lost 25% of them. How many games did they win?
A. 5 games.
B. 10 games.
C. 15 games.
D. 17 games.

ANSWER = C

24. In the first promotional exam that Captain Jones took there were 100 questions and he had 80 questions correct. In the next exam having the same number of questions he had 10% more correct than he did in the first exam. How many did he have correct on the second exam?
A. 88.
B. 90.
C. 92.
D. None of the above.

ANSWER = A

25. Fireman Carlson bought a flashlight for $7.50. After using it for two shifts he sold it for a 20% loss. At what price did he sell the flashlight?
A. $5.00.   C. $6.00.
B. $5.50.   D. $6.50.

ANSWER = C

26. The Firefighter Auxiliary earned a $120.00 commission from the sale of magazine subscriptions. This amount was equal to 25% commission on all sales. How many dollars worth of magazine subscriptions did they sell?
A. $240.00.   C. $340.00.
B. $480.00.   D. $680.00.

ANSWER = B

27. The State Forestry Division of Fire map has a scale where 1 inch = 100 miles. How many miles are represented by 2 1/4 inches on the map?
A. 225 miles.   C. 250 miles.
B. 220 miles.   D. 200 miles.

ANSWER = A

28. Fireman Jone's salary was increased from $1,000.00 per pay period to $1,200.00 per pay period. What per cent of increase is this?
    A. 5% increase.
    B. 10% increase.
    C. 15% increase.
    D. 20% increase.

    ANSWER = D

29. Last year the City Fire Department responded to 8,875 incidents. This year they responded to 9,230 incidents. What per cent of increase is this?
    A. 4%.
    B. 5%.
    C. 6%.
    D. None of the above.

    ANSWER = A

30. Fireman Paramedic Cannon wanted an early relief from Fireman Paramedic Roberts at 0500 hours. Roberts was 1 hour and 40 minutes late. At what time did he arrive?
    A. 0540 hours.
    B. 0630 hours.
    C. 0634 hours.
    D. None of the above.

    ANSWER = D

31. Firefighter Simpson is 5 feet, 9 inches tall and Captain Sutton is 67 inches tall. Which man is taller.
    A. Firefighter Simpson.
    B. Captain Sutton.
    C. They are the same height.
    D. None of the above.

    ANSWER = A

32. How much would a Firefighter pay for 5 "Pen-Lights" at a rate of $6.00 per dozen?
    A. $1.50.
    B. $2.00.
    C. $2.50.
    D. $3.00.

    ANSWER = C

33. If 14% of the Firefighters live within the City, what per cent live outside of the City.
    A. 86%.
    B. .14%.
    C. .86%.
    D. None of the above.

    ANSWER = A

34. Firefighter Smith had three Library books that were 4 days overdue. His fine was $.12 per day for each book. How much was his total fine?
   A. $1.24.
   B. $1.34.
   C. $1.43.
   D. $1.44.

   ANSWER = D

35. A year's subscription of "Firefighters" monthly magazine cost $19.95 per year. If a single copy cost $2.00, how much will a Firefighter save by taking a year's subscription instead of buying a single copy each month?
   A. $3.05.
   B. $4.05.
   C. $5.03.
   D. $5.04.

   ANSWER = B

36. What is the difference in cost to a firefighter between a flashlight listed at $50.00 less 10% and one listed at $45.00 less 20% ?
   A. $1.00.
   B. $1.50.
   C. $2.00.
   D. $2.50.

   ANSWER = A

37. Firefighter Wendl borrowed $4000.00 at 14%. How much additional money can he borrow at 16% if his total interest is not to exceed $1250.00 a year?
   A. $ 800.00.
   B. $ 900.00.
   C. $2000.00.
   D. $4000.00.

   ANSWER = D

38. Fire Engineer Smith wishes to construct a storage box 12 feet long to keep spare equipment in. If each piece of equipment requires 4 square feet of space, how wide should he construct the box ?
   A. 6 1/2 feet.
   B. 6 feet 8 inches.
   C. 20 feet.
   D. 80 feet

   ANSWER = B

39. How much longer does it take a ladder truck to travel one mile at 20 miles per hour than at 30 miles per hour?
    A. 1 minute.
    B. 10 minutes.
    C. 20 minutes.
    D. 40 minutes.

    ANSWER = A

40. Fire Captain Jones stated to Firefighter Smith that light travels approximately 186,000 miles a second and that the sun is 933 million miles away from the earth. Firefighter Smith asked Captain Jones if he could figure out how long it would take a ray of light to travel from the sun to the earth. Captain Jones said that he could figure it out to the nearest minute. What would be the correct time?
    A. 2 minutes.
    B. 5 minutes.
    C. 8 minutes.
    D. 58 minutes.
    E. None of the above.

    ANSWER = C

SECTION 8

FIRST AID / EMT TECHNIQUES

EXPLANATION

Many entrance level Firefighter exams will include some questions relating to first aid. This portion of the exam is designed to measure the candidates knowledge of the basic principles and procedures of first aid.

When encountering the first aid portion of the exam: read each question carefully so as not to misinterpret the meaning of the question. Then try to reason out the answer. If you are not sure of the answer, eliminate the obvious incorrect answers. Then choose the best answer.

Never choose an answer without reading all of the choices. Also don't instantly exclude a question in which you have noticed unfamiliar technical terminology. After reading the complete question you might discover that you recognize the answer after all.

Many Fire Departments require that prospective entrance level Firefighter candidates be certified as: **EMERGENCY MEDICAL TECHNICIAN - I (EMT-I)**.

Included in this section is information relating to **EMT-I** along with typical questions relating to the subject.

FIRST AID INFORMATION

First Aid is the immediate and temporary care given the victim of an accident or sudden illness until the services of a physician can be obtained.

First Aid commences with the steadying effect upon the stricken person when he/she realizes that competent hands will help him.

The good First Aider deals with the whole situation, the person and the injury. He/she knows what not to do as well as what to do. He/she confines procedures to what is necessary, recalling that the handling of the injured be kept to a minimum.

General directions for First Aid:
1. Give urgently necessary First Aid for the three injuries that must be treated immediately to avoid the loss of life: severe bleeding, stoppage of breathing, and poisoning.
2. Keep the victim lying down: protect him/her from unnecessary manipulation and disturbance. Do not heat the patient but keep body temperatures from falling.

3. Check for injuries: Have a reason for what you do. Check certain parts of the body for injuries because of clues given you by the patient, witnesses, or other circumstances. Avoid twisting, bending, or shaking of body parts.
4. Plan what to do: Telephone for further assistance from a Physician or Paramedic/EMT, 911, ETC!
5. Carry out the indicated First Aid.
6. Additional task to perform: find all injuries, treat minor as well as major injuries. Reassure the patient. ETC!

FIRST AID TOPICS TO KNOW

1. Wounds
2. Shock.
3. Artificial Respiration.
4. CPR.
5. Poisoning by mouth.
6. Injuries to bones, joints and muscles.
7. Burns.
8. Ill effects of heat and cold.
9. Common emergencies.
10. Transportation.
11. The human body.
12. Special wounds.

FIRST AID QUESTIONS

1. Of the following choices, which is the one that most dwelling accidents result from:
   A. Burns.
   B. Falls.
   C. Poisons.
   D. Suffocation.

   ANSWER = B

2. While in an enclosed area, death may occur from an automobile engine running. Of the following choices, which one is the most likely cause:
   A. Carbon monoxide poisoning.
   B. Carbon dioxide in the air.
   C. Excess humidity.
   D. Suffocation.

   ANSWER = A

3. Arterial pressure points are:
   A. Deep seated and require great pressure.
   B. Best located by taking a pulse.
   C. Close to bones near the surface of the body.
   D. Used to cut off all blood circulation.

   ANSWER = C

4. In the treatment of heat exhaustion, of the following which should be the first task:
A. Have patient lie down keeping the head elevated.
B. Keeping the body covered and warm.
C. Treat for shock.
D. Move the patient to an area with cool circulating air.

ANSWER = D

5. Of the following, which are the best examples of the symptoms of heat exhaustion:
A. Pale, clammy skin, low temperature, weak pulse.
B. Headache, red face, unconsciousness.
C. Abdominal cramps, red skin, profuse sweating.
D. Rapid strong pulse, dry skin, high temperature.

ANSWER = A

6. The heart of a NORMAL male adult, at rest, beats approximately:
A. 36 times a minute.
B. 70 times a minute.
C. 96 times a minute.
D. 106 times a minute.

ANSWER = B

7. Of the following methods, which is the best for reducing swelling:
A. Cold packs.
B. Hot packs.
C. Elastic bandage.
D. Warm whirlpool.

ANSWER = A

8. Of the following terms, which best describes the result of a direct blow upon a muscle:
A. Contusion.
B. Fracture.
C. Sprain.
D. Strain

ANSWER = A

9. If a patient suffers a compound fracture of the leg, which of the following choices would be the most likely involved:
A. Clavicle.
B. Sternum.
C. Tibia or fibula.
D. Radius or ulna.

ANSWER = C

10. Of the following choices, which choice would not be a symptom of shock:
    A. Cold clammy skin.
    B. Flushed face.
    C. Weak pulse.
    D. Feeling of weakness.

    ANSWER = B

11. Of the following illnesses, which one would present: fever, chills, inflamed eyelids, running nose, and a cough:
    A. Chicken pox.
    B. Tuberculosis.
    C. Measles.
    D. Scarlet fever.

    ANSWER = C

12. Of the following choices of First Aid for a victim of a head injury, which is correct:
    A. Have patient lying down with his/her head level or lower than the rest of body.
    B. Have patient lying down with his/her head raised somewhat.
    C. Have patient sit with head bent down.
    D. Have patient continue to move his/her head in motion to increase circulation.

    ANSWER = B

13. As a witness to a man falling off the roof of a one story home and becomes unconscious, of the following what should you do:
    A. Hold his head up and give him a stimulant.
    B. Move him inside the house.
    C. Call for help and do not move the man.
    D. Walk him around until he awakens.

    ANSWER = C

14. Of the following, which is the correct First Aid procedure for chemical burns:
    A. Apply baby oil solution to the burn.
    B. Apply a dry dressing.
    C. Dry dressing after through flushing with water.
    D. Flush with water and leave uncovered.

    ANSWER = C

15. Of the following First Aid treatments, which should you administer to a patient suffering from shock:
    A. Cover patient and keep warm.
    B. Cover with cold towels.
    C. Keep patient alert.
    D. Give patient caffeine.

    ANSWER = A

16. Of the following problems, which is the one that artificial respiration would **NOT** be the proper First Aid treatment:
    A. Asphyxiation.
    B. Electrical shock.
    C. Drowning.
    D. Internal bleeding.

    ANSWER = D

17. Patients of sunstroke and heat exhaustion are similar in that both patients:
    A. Have hot dry skin.
    B. Should lie down with their heads raised.
    C. Have been exposed to heat.
    D. Should be given caffeine.

    ANSWER = C

18. In the First Aid treatment of a victim of a poisoning, the main reason for giving the patient water is to:
    A. Dilute the poison.
    B. Calm the patient.
    C. Avert choking.
    D. Give the patient energy.

    ANSWER = A

19. In First Aid, the technique of artificial respiration is used to:
    A. Create pressure on the heart.
    B. Circulate blood.
    C. Keep the patient war.
    D. Force air into the lungs.

    ANSWER = D

20. When treating a patient of sever shock, it is vital that the patient receives:
    A. Cool water and a tranquilizer.
    B. Warm coffee and is moved.
    C. Warmth and keeps head lower than rest of body.
    D. Tranquilizers and is seated.

    ANSWER = C

21. Of the following, the best reason that the person administering First Aid should not disturb blood clots, is that:
    A. Blood clots can cause infection.
    B. The blood of most people clots very easily.
    C. Clotted blood does not indicate whether a vein or artery has been involved.
    D. Blood clots stop bleeding.

    ANSWER = D

22. While searching an area after an earthquake you discover a victim with a leg injury, bleeding severely and in a lot of pain. In this situation the first action in regards to First Aid should be:
A. Immobilize the patients leg.
B. Cover the patient with a blanket.
C. Stop the bleeding.
D. Comfort the patient.

ANSWER = C

23. At the scene of an incident where there are several patients, which of the following problems should be treated first, with First Aid:
A. Unconscious patient.
B. Patient with sever bleeding.
C. Patient that is moaning in pain.
D. Patient that is vomiting.

ANSWER = B

24. Of the following, what treatment should be given to a patient that has a laceration of about two inches long on the right leg that is bleeding moderately:
A. Apply pressure directly to the wound.
B. Apply a tourniquet between the wound and the heart.
C. Apply pressure to the pressure point directly above the wound.
D. Allow the wound to bleed so as to avoid infection.

ANSWER = A

25. Of the following choices, which is the best First Aid treatment for a patient having a seizure:
A. Give patient water.
B. Prevent patient from injuring himself.
C. Hold patient down.
D. Hold patients mouth open.

ANSWER = B

EMT-I INFORMATION

Emergency medical technicians have the greatest occasion of any portion of a community to relieve human distress at the scene of an accident or sudden illness. With the proper training and experience, supplies and equipment to meet the needs of the sick and injured, the Emergency Medical Technician (EMT-I) is equipped to assist as the most useful lay participant of the emergency medical care team outside of the hospital setting.

Emergency medical care is a team effort, provided by a team consisting of those among the general public who are trained in first aid, the EMT-I, hospital emergency room personnel, hospital administration, health personnel, emergency services, and the municipality.

Wherever injury or sudden illness strikes, there will rarely be trained medical personnel on scene to administer initial care. Usually the people on hand will be limited in their ability to maintain patients with life threatening conditions until the proper assistance arrives. The limitations can be because of training or even in cases where they have had the training, the limitation can be due to the fact that they do not have the needed equipment, supplies, or means of transportation that may be necessary to save life and safely transport a patient to the hospital.

It is necessary that the qualified EMT-I be able to asses the caliber and effectiveness of first aid already given by those of less training and to take direct charge without delay and tactfully, also to enlist the assistance of others as needed.

EMT-I TOPICS TO KNOW

1. Legal responsibilities.
2. Anatomy and physiology.
3. Diagnostic signs and triage.
4. Basic life support:
   The respiratory system.
   Injuries to the chest.
   The circulatory system.
   Bleeding and control of bleeding.
   Shock.
   Basic life support.
   Oxygen therapy.
5. The musculoskeletal system:
   The skin.
   The muscular system.
   Soft tissue injuries.
   The skeletal system.
   Fractures, dislocations, and sprains.
   General principles of splinting and bandaging.
   Fractures and dislocations of:
      Shoulder and upper extremity.
      Hip, pelvis, and lower extremity.
      Spine.
6. The head and nervous system:
   The nervous system.
   Injuries of the skull and brain.
   The eye.
   Injuries of the eye, face, and throat.
7. The abdomen and genitourinary system:
   The abdomen and digestive system.
   Injuries of the abdomen.
   The acute abdomen.
   The genitourinary system.
   Injuries of the genitourinary system.
8. Medical emergencies:
   Heart attack.
   Stroke.
   Diabetes mellitus.
   Dyspnea.
   Unconscious states.
   Communicable disease.
9. Childbirth and special pediatric problems:
   Childbirth.
   Pediatric emergencies and special problems.
10. Mental health problems:
    The disturbed and unruly patient.
    Alcohol and drug abuse.

11. Environmental injuries:
    Heat exposure.
    Cold exposure.
    Radiation exposure.
    Electrical hazards.
    Water hazards.
    Poisons, stings, and bites.

12. Emergency vehicles and equipment:
    Patient handling.
    Extrication.
    Equipment maintenance.
    Emergency driving.
    Traffic control.
    Accident scene.

14. Communications:
    Records.
    Reports.
    Communications.

15. Medical terminology.

TYPICAL EMT-I QUESTIONS

1. On a hot humid day an elderly man is found unconscious on a park bench. His skin feels hot and dry, and his pulse is strong and regular. You suspect:
   A. Insulin shock.
   B. Heat exhaustion.
   C. Heat stroke.
   D. Epilepsy.

   ANSWER = C

2. Paradoxical respirations are indicative of a patient with a:
   A. Flail chest.
   B. Cardiac tamponade.
   C. Ruptured spleen.
   D. Pneumothorax.

   ANSWER = A

3. The signs of pericardial tamponade include:
   A. Distended neck veins.
   B. Narrowing pulse pressure.
   C. Faint heart sounds.
   D. All of the above.

   ANSWER = D

4. Respiration rate and depth are often indicative of disease or injury. Kussmaul respirations are often a sign of:
   A. Hypertension.
   B. Hyperglycemia.
   C. Emphysema.
   D. Cardiac arrest.

   ANSWER = B

5. Trauma to the left upper quadrant from a steering wheel would most likely involve the:
   A. Bladder.
   B. Kidney.
   C. Liver.
   D. Spleen.

   ANSWER = D

6. In resuscitation of a newborn baby, the first step should be to:
A. Begin CPR.
B. Give four rapid puffs of air.
C. Suction the mouth and nose to remove fluids.
D. Tilt the head all the way back.

ANSWER = C

7. You witness a person that has been splashed with a caustic chemical. Of the following, what is the first thing that you should do:
A. Wipe the caustic chemical off.
B. Submerge the person in warm water.
C. Flush the affected area with copious amounts of water.
D. Neutralize the caustic chemical with phenol acid.

ANSWER = C

8. Of the following, which one is not classified as a narcotic:
A. Codine.
B. Darvon.
C. Morphine.
D. Valium.

ANSWER = A

9. A chronic lung disease which results in loss of elasticity of the aveoli is:
A. Asthma.
B. Bronchitis.
C. Emphysema.
D. Pneumonitis.

ANSWER = C

10. A barrel hoop test is used to locate fractures of the ribs and:
A. Femur.
B. Pelvis.
C. Shoulder.
D. Spine.

ANSWER = B

11. A fractured hip will normally cause the foot to:
A. Rotate out, lateral.
B. Rotate in, medially.
C. Loose sensation.
D. Spasm uncontrollably.

ANSWER = A

12. Of the following, in which condition may vomiting be induced:
    A. Ingestion of strong acid or alkaline/caustic substances.
    B. Ingestion of petroleum products.
    C. In a patient with no gag reflex/unconscious patient.
    D. None of the above.

    ANSWER = D

13. Of the following patients, which one should not be positioned with the upper body elevated:
    A. Difficulty breathing, good airway.
    B. Head injury, not unconscious.
    C. Nosebleed.
    D. Shock.

    ANSWER = D

14. Of the following, which medical emergency should be transported the most rapidly:
    A. Anaphylaxis.
    B. Heat exhaustion.
    C. Spine injury.
    D. Stroke patient.

    ANSWER = A

15. **REFERRED PAIN** to the shoulder is **USUALLY** associated to:
    A. Acute abdomen.
    B. Cerebral disorder.
    C. Pelvic fracture.
    D. Pulmonary disorder.

    ANSWER = A

16. A person suffering from diabetic coma would most likely appear with which of the following symptoms:
    A. Normal skin, bounding pulse, rapid deep respirations.
    B. Warm dry skin, rapid weak pulse, air hunger.
    C. Cool dry skin, slow weak pulse, normal respirations.
    D. Pale moist skin, full rapid pulse, normal respirations.

    ANSWER = B

17. Of the following, which breathing problem is manifested by: anxiousness, decreasing B/P, possible syncope and carpalpedal spasms:
    A. Asthma.
    B. Emphysema.
    C. Hyperventilation.
    D. Pulmonary edema.

    ANSWER = C

18. Of the following, the first consideration with a near drowning patient would be:
    A. Getting the patient out of the water.
    B. Getting the water out of the patients lungs.
    C. Clearing the airway.
    D. C-Spine precautions.

    ANSWER = C

19. Of the following, which is the most desirable technique of immobilization for a fractured humerus:
    A. Elevation.
    B. Sling only.
    C. Traction.
    D. Sling and swath.

    ANSWER = D

20. Of the following splinting methods, which is the best one to use for a fractured femur:
    A. Air splint.
    B. Traction splint.
    C. Cardboard splint.
    D. Vacuum splint.

    ANSWER = B

21. A construction worker falls from the roof of a house. As an EMT-I you arrive on the scene and he is unconscious, has unequal pupils and widening pulse pressure. This would indicate:
    A. Increased intracranial pressure, probably from bleeding on one side of the brain.
    B. Broken neck.
    C. Concussion.
    D. Tamponade.

    ANSWER = A

22. A gang member has been stabbed in the left side of his chest. As an EMT-I you arrive on the scene and the knife is still in his chest. You should
    A. Have the victim cough, then pull the knife out.
    B. Stabilize the knife with bulky dressings and administer oxygen.
    C. Apply pressure directly on the knife to prevent further loss of blood.
    D. Immediately remove the knife and apply a pressure dressing.

    ANSWER = B

23. As a firefighter, EMT-I, you respond to a house fire and find one of the occupants has burns covering the entire front of his chest and abdomen and both arms, front and back. Of the following, which is most nearly the estimated area of burn:
    A. 36%.
    B. 40%.
    C. 44%.
    D. 54%.

    ANSWER = A

24. As an EMT-I you respond to a home at 2:00 am and find a three year old female having difficulty breathing. She is extremely anxious and has a seal-like cough. Of the following which is most likely her problem:
    A. Asthma.
    B. Epiglottitis.
    C. Croup.
    D. Meningitis.

    ANSWER = C

25. As an EMT-I you respond to the scene of an automobile accident and find that the victim is conscious, has a stiff neck and has no sensation or movement in his lower extremities. Of the following examples, you should:
    A. Remove patient from car and transport.
    B. Remove patient from the car, apply cervical collar and transport.
    C. Apply cervical collar, remove patient from the car and transport.
    D. Stabilize the entire spine with a backboard and cervical collar before removing the patient from the car.

    ANSWER = D

SECTION 9 PROGRESSIONS

EXPLANATION

There are several kinds of progressions including alphabetical and numerical. Progressions may have more than one or two patterns. these types of test are used as a method to find the candidates ability to answer questions he/she has not previously been trained. Also to test the candidates ability to work and reason independently of supervision, and the capacity for logical reasoning.

When encountering the portion of the exam that introduces progressions: Study each question carefully and base your answers on the data given with each question. Remain calm even if the progression looks hopelessly foreign or complicated at first. Once you have studied the progression you will probably be able to figure out a solution. Be accurate, sometimes the printed progression or figures are small and hard to read. Make every effort to take all data into consideration before answering the question.

Don't ignore details, but don't spend too much time on any one question.

When encountering the portion of the exam that introduces alphabetical progressions: you should make use of scratch paper and print the letters of the alphabet in the correct order on one line. Below this line, print the alphabet in reverse order. Doing this will increase your speed in taking the exam.

When encountering the portion of the exam that introduces numerical progressions: be aware that they are not usually the same as alphabetical progressions and are not solved in the same way. The patterns are usually variants of every number, every other number, or every third number. In a numbered progression, each is plus, minus, multiplied by, or divided by a number from the last number in the progression.

ALPHABETICAL PROGRESSIONS

INSERT THE CORRECT LETTER THAT WILL COMPLETE THE FOLLOWING ALPHABETICAL PROGRESSIONS:

EXAMPLES:

#1. A B C D __
ANSWER = E

#2. A C E G __
ANSWER = I

#3. L O P S T __
ANSWER = W

#4. AA CC EE GG ___
ANSWER = HH

1. T R P N __
ANSWER = L

2. K M O Q __
ANSWER = S

3. A D G J __
ANSWER = M

4. M J G D __
ANSWER = N

5. C E H L __
ANSWER = Q

6. V T R P __
ANSWER = N

7. B E J Q __
ANSWER = Z

8. A D G J __
ANSWER = M

9. G H I J K __
ANSWER = L

10. AA DD GG JJ ___
ANSWER = MM

11. CC EE GG II ___
ANSWER = KK

12. A C B D C E __
ANSWER = D

13. Z W T __
ANSWER = Q

14. XX UU RR ___
ANSWER = OO

15. D F E H G __
ANSWER = K

NUMERICAL PROGRESSION

IN THE FOLLOWING NUMBER PROGRESSIONS, ADD THE CORRECT NUMBER THAT SHOULD BE PLACED AT THE END OF EACH SERIES:

EXAMPLES:

#1. 1 2 3 4 5 __   #2. 5 10 15 20 25 __
     ANSWER = 6   ANSWER = 30

1. 1 3 5 7 9 11 13 __
   ANSWER = 15

2. 1 2 6 7 21 22 66 __
   ANSWER = 67

3. 2 2 3 3 5 5 8 __
   ANSWER = 8

4. 7 10 8 12 9 12 10 __
   ANSWER = 14

5. 20 19 21 19 22 19 23 __
   ANSWER = 19

6. 2 4 8 16 32 64 128 __
   ANSWER = 256

7. 4 8 9 13 14 18 19 __
   ANSWER = 23

8. 4 6 12 14 28 30 60 62 __
   ANSWER = 62

9. 4 4 0 5 5 1 6 __
   ANSWER = 6

10. 0 1 3 6 10 15 21 __
    ANSWER = 28

11. 120 124 62 64 32 36 18 22 __
    ANSWER = 22

220

12. 5 10 7 14 11 22 19 __
    ANSWER = 38
13. 20 10 12 6 4 2 4 __
    ANSWER = 2
14. 25 27 30 15 17 20 10 __
    ANSWER = 12
15. 30 23 17 12 8 5 __
    ANSWER = 3
16. 67 59 64 16 48 40 __
    ANSWER = 45
17. 2 6 18 54 162 __
    ANSWER = 486
18. 6 8 10 8 10 12 10 __
    ANSWER = 12
19. 5 6 7 8 16 17 34 __
    ANSWER = 35
20. 96 24 28 30 10 13 14 __
    ANSWER = 7
21. 3 6 11 17 26 37 50 __
    ANSWER = 65

22. 9 12 15 18 21 24 __
    ANSWER = 27

23. 3 8 16 21 63 68 136 __
    ANSWER = 141

24. 9 10 8 24 6 7 5 __
    ANSWER = 15

25. 2 6 12 20 30 42 56 __
    ANSWER = 72

SECTION 10

MATCHING FORMS PATTERN ANALYSIS

EXPLANATION

There are several kinds of matching forms and. These types of test are used as a method to find the candidates ability to answer questions he/she has not previously been trained. Also to test the candidates ability to work and reason independently of supervision, and the capacity for logical reasoning.

When encountering the portion of the exam that introduces forms, cubes, blocks, and figures: Study each question carefully and base your answers on the data given with each question. Remain calm even if the diagram looks hopelessly foreign or complicated at first. Once you have studied the form you will probably be able to figure out a solution. Be accurate, sometimes the printed figures/drawings are small and hard to read. Make every effort to take all data into consideration before answering the question.

Don't ignore details, but don't spend too much time on any one question.

When encountering the portion of the exam that deals with forms/cubes/blocks/figures: be aware that these habitually consist of two or more symbols which are related in some way: specific type of change of form, the addition or subtraction of a portion of design, the change in direction of a portion of the design, etc. The candidate is asked to deduce the next symbol in the series of symbols.

CUBES:

IN ALL OF THE FOLLOWING QUESTIONS PERTAINING TO CUBES/BLOCKS, IN EACH PARTICULAR DRAWING THE CUBES/BLOCKS ARE OF THE IDENTICAL DIMENSIONS AND SHAPE; ALTHOUGH THE DIMENSIONS AND SHAPE MAY VARY FROM ONE QUESTION TO ANOTHER. IF NECESSARY, DON'T FORGET TO COUNT UNSEEN BOXES NEEDED TO SUPPORT DRAWINGS:

1. How many blocks are there in the drawing below?

   A. 2 Blocks.
   B. 3 Blocks.
   C. 4 Blocks.
   D. 5 Blocks.

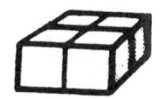

   ANSWER = C

2. How many blocks are there in the drawing below?

   A. 6 Blocks.
   B. 8 Blocks.
   C. 9 Blocks.
   D. 10 Blocks.

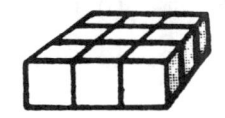

   ANSWER = C

3. How many blocks are there in the drawing below?

   A. 8 Blocks.
   B. 10 Blocks.
   C. 12 Blocks.
   D. 16 Blocks.

   ANSWER = C

4. How many blocks are there in the drawing below?

   A. 12 Blocks.
   B. 15 Blocks.
   C. 16 Blocks.
   D. 18 Blocks.

   ANSWER = B

5. How many blocks are there in the drawing below?

   A. 36 Blocks.
   B. 40 Blocks.
   C. 42 Blocks.
   D. 44 Blocks.

   ANSWER = B

6. How many blocks are there in the drawing below?

   A. 22 Blocks.
   B. 24 Blocks.
   C. 26 Blocks.
   D. 27 Blocks.

   ANSWER = A

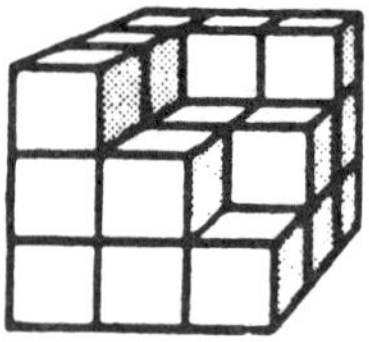

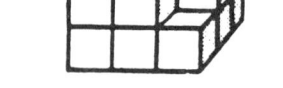

7. How many blocks are there in the drawing below?

   A. 9 Blocks.
   B. 15 Blocks.
   C. 16 Blocks.
   D. 18 Blocks.

   ANSWER = D

8. How many blocks are there in the drawing below?

   A. 10 Blocks.
   B. 12 Blocks.
   C. 13 Blocks.
   D. 15 Blocks.

   ANSWER = D

9. How many blocks are there in the drawing below?

   A. 15 Blocks.
   B. 18 Blocks.
   C. 19 Blocks.
   D. 21 Blocks.

   ANSWER = C

10. How many blocks are there in the drawing below?

    A. 6 Blocks.
    B. 7 Blocks.
    C. 8 Blocks.
    D. 9 Blocks.

    ANSWER = A

11. How many blocks are there in the drawing below?

    A. 5 Blocks.
    B. 6 Blocks.
    C. 7 Blocks.
    D. 8 Blocks.

    ANSWER = C

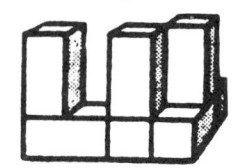

12. How many blocks are there in the drawing below?

   A. 18 Blocks.
   B. 20 Blocks.
   C. 21 Blocks.
   D. 24 Blocks.

   ANSWER = B

13. How many blocks are there in the drawing below?

   A. 5 Blocks.
   B. 6 Blocks.
   C. 7 Blocks.
   D. 8 Blocks.

   ANSWER = B

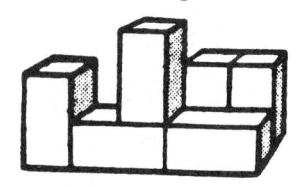

14. How many blocks are there in the drawing below?

   A. 15 Blocks.
   B. 16 Blocks.
   C. 17 Blocks.
   D. 18 Blocks.

   ANSWER = A

15. How many blocks are there in the drawing below?

   A. 8 Blocks.
   B. 10 Blocks.
   C. 12 Blocks.
   D. 15 Blocks.

   ANSWER = D

16. How many blocks are there in the drawing below?

   A. 9 Blocks.
   B. 10 Blocks.
   C. 12 Blocks.
   D. 13 Blocks.

   ANSWER = C

17. How many blocks are there in the drawing below?

   A. 10 Blocks.
   B. 11 Blocks.
   C. 12 Blocks.
   D. 13 Blocks.

   ANSWER = C

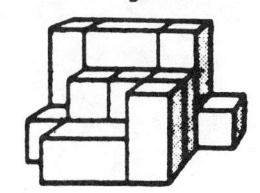

18. How many blocks are there in the drawing below?

    A. 7 Blocks.
    B. 8 Blocks.
    C. 9 Blocks.
    D. 10 Blocks.

    ANSWER = B

19. How many blocks are there in the drawing below?

    A. 13 Blocks.
    B. 14 Blocks.
    C. 15 Blocks.
    D. 16 Blocks.

    ANSWER = D

20. How many blocks are there in the drawing below?

    A. 27 Blocks.
    B. 28 Blocks.
    C. 29 Blocks.
    D. 30 Blocks.

    ANSWER = D

21. How many blocks are there in the drawing below?

    A. 23 Blocks.
    B. 25 Blocks.
    C. 27 Blocks.
    D. 28 Blocks.

    ANSWER = B

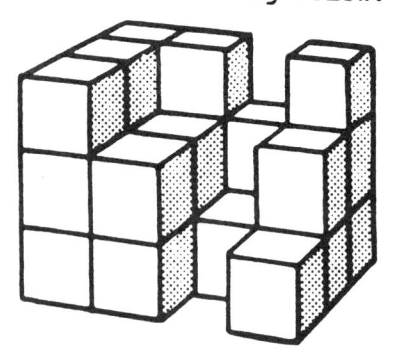

22. How many blocks are there in the drawing below?

   A. 22 Blocks.
   B. 17 Blocks.
   C. 21 Blocks.
   D. 18 Blocks.

   ANSWER = A

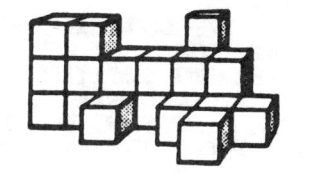

23. How many blocks are there in the drawing below?

   A. 7 Blocks.
   B. 8 Blocks.
   C. 9 Blocks.
   D. 10 Blocks.

   ANSWER = B

24. How many blocks are there in the drawing below?

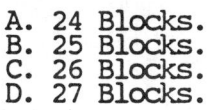

   A. 24 Blocks.
   B. 25 Blocks.
   C. 26 Blocks.
   D. 27 Blocks.

   ANSWER = D

25. How many blocks are there in the drawing below?

   A. 25 Blocks.
   B. 33 Blocks.
   C. 35 Blocks.
   D. 36 Blocks.

   ANSWER = B

MATCHING FORMS, FIGURES ETC.

IN THE FOLLOWING DRAWINGS, FIND THE CUBE AT THE RIGHT THAT COULD BE THE ORIGINAL CUBE ROTATED INTO A DIFFERENT POSITION. AMONG THE THREE DRAWINGS AT THE RIGHT, ONE DRAWING WILL BE THE ORIGINAL CUBE TURNED TO A DIFFERENT POSITION. CUBES MAY BE ROTATED TO ANY POSITION. THERE IS A DIFFERENT DESIGN ON EACH OF THE SIX SURFACES OF EACH CUBE.

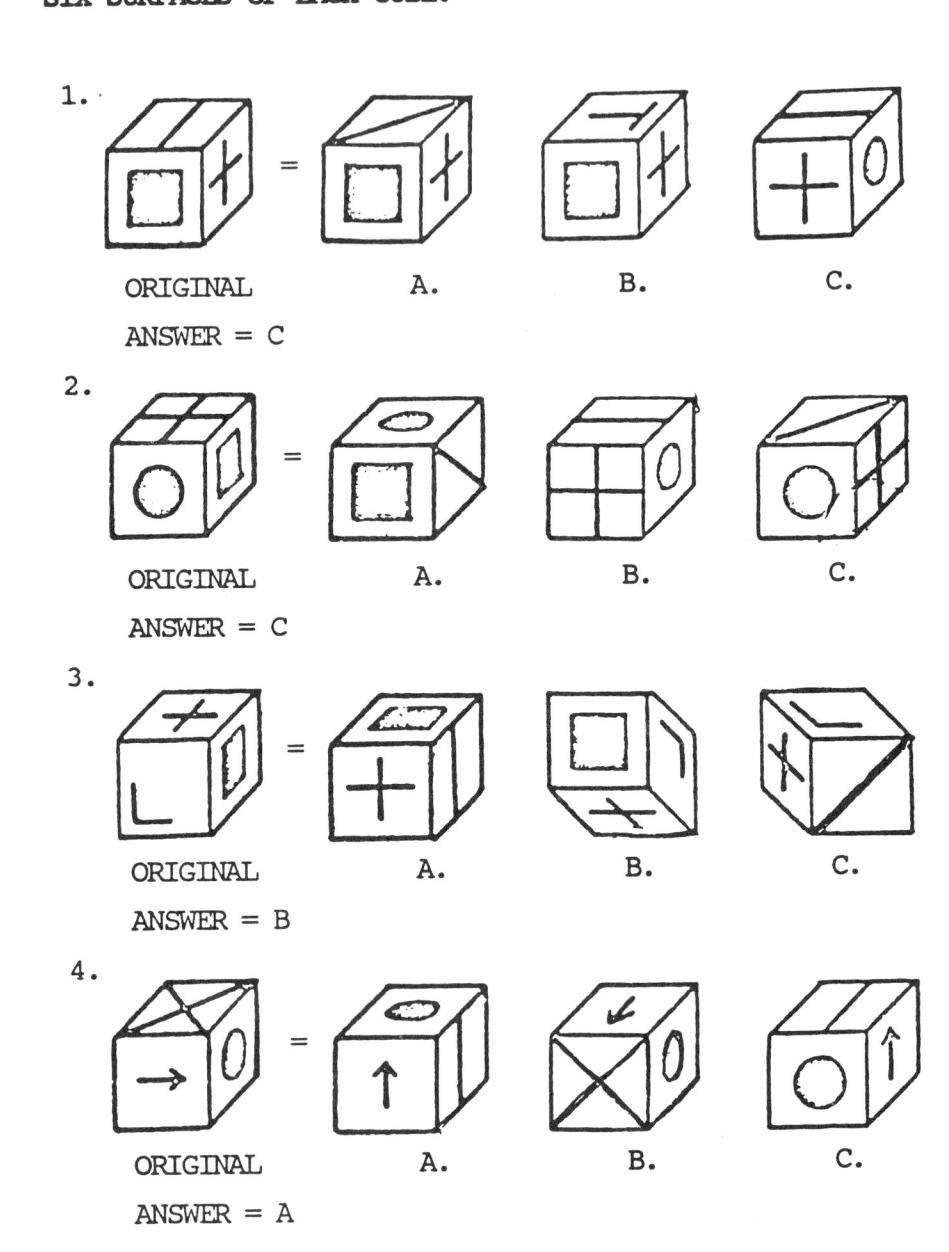

1.

ORIGINAL   A.   B.   C.

ANSWER = C

2.

ORIGINAL   A.   B.   C.

ANSWER = C

3.

ORIGINAL   A.   B.   C.

ANSWER = B

4.

ORIGINAL   A.   B.   C.

ANSWER = A

5.

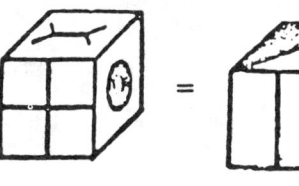

ORIGINAL   A.   B.   C.

ANSWER = A

6.

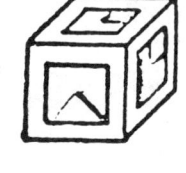

ORIGINAL   A.   B.   C.

ANSWER = B

7.

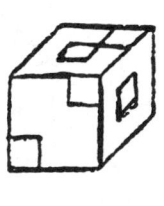

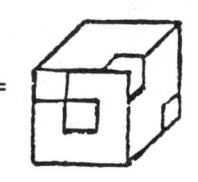

ORIGINAL   A.   B.   C.

ANSWER = B

8.

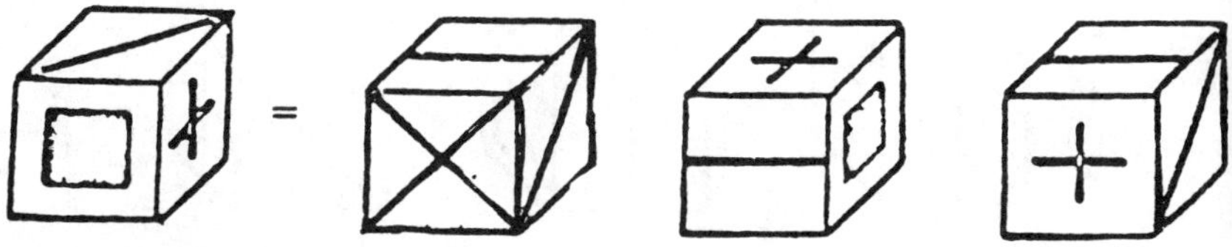

ORIGINAL   A.   B.   C.

ANSWER = A

9.

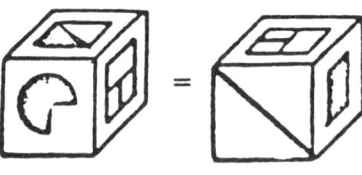

ORIGINAL   A.   B.   C.

ANSWER = B

10.

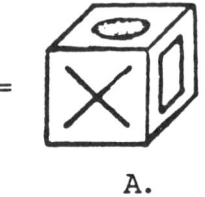

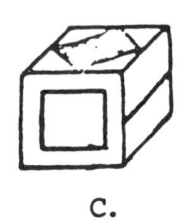

ORIGINAL   A.   B.   C.

ANSWER = C

11.

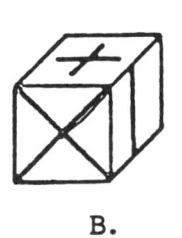

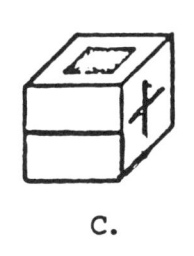

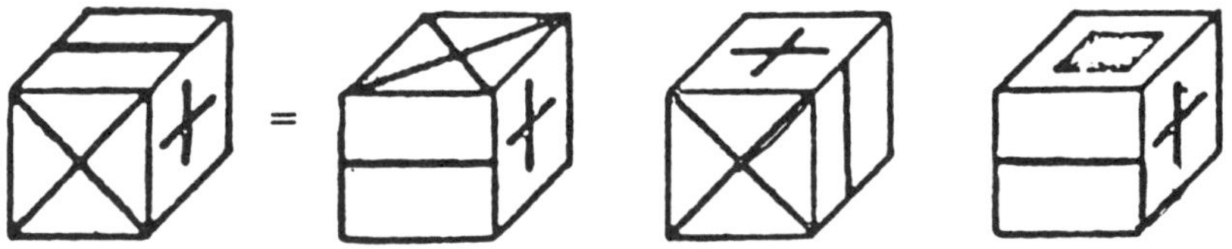

ORIGINAL   A.   B.   C.

ANSWER = C

12.

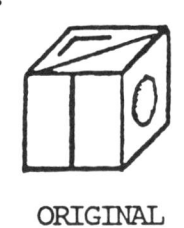

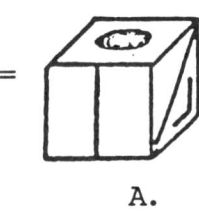

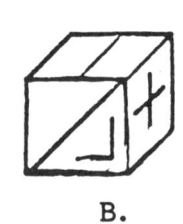

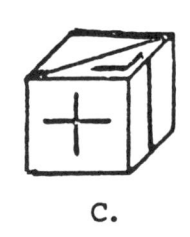

ORIGINAL   A.   B.   C.

ANSWER = B

13.

ORIGINAL   A.   B.   C.

ANSWER = A

14.

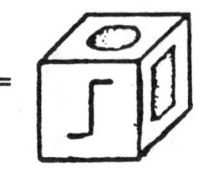

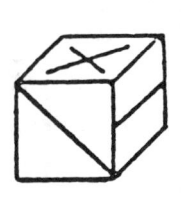

ORIGINAL   A.   B.   C.

ANSWER = C

15.

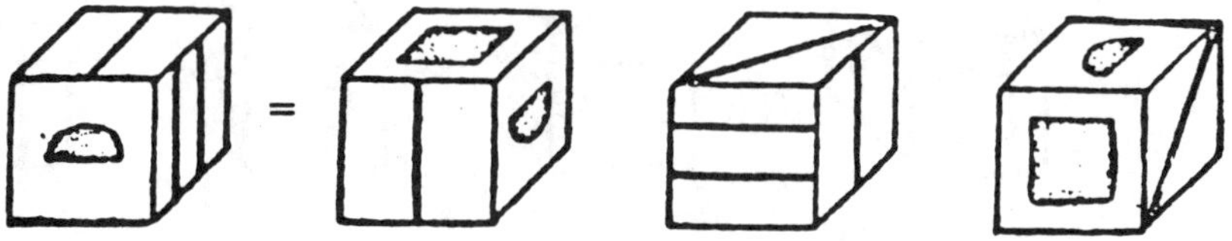

ORIGINAL   A.   B.   C.

ANSWER = C

16.

ORIGINAL   A.   B.   C.

ANSWER = B

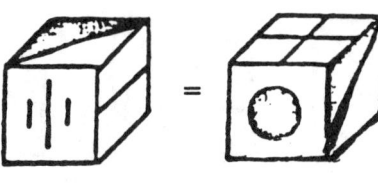

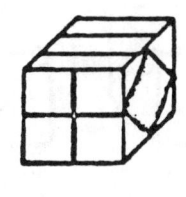

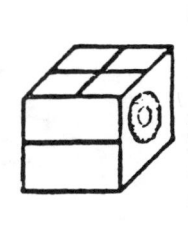

17.

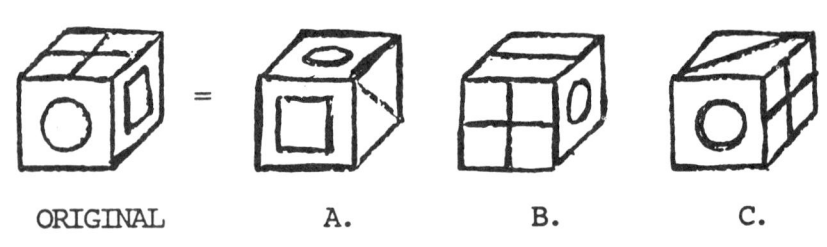

ORIGINAL   A.   B.   C.
ANSWER = C

18.

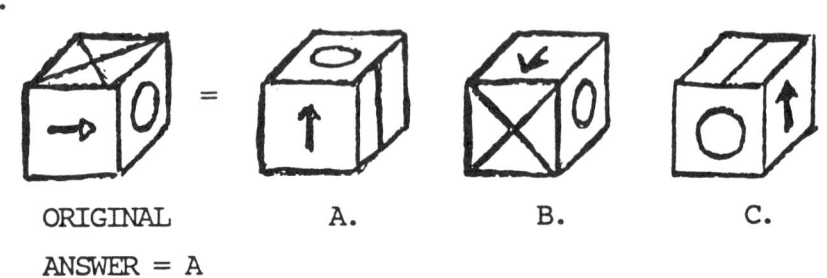

ORIGINAL   A.   B.   C.
ANSWER = A

19.

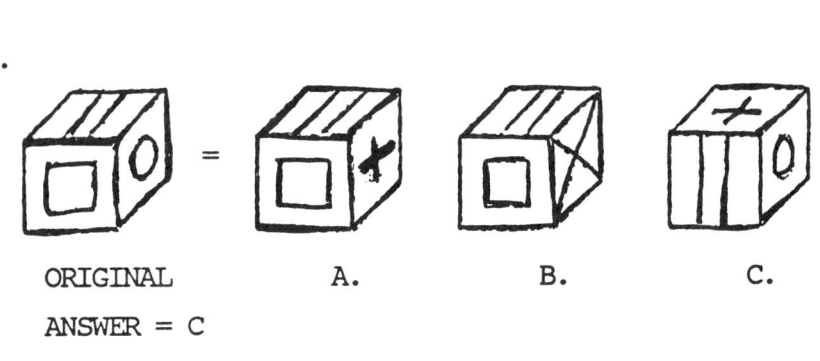

ORIGINAL   A.   B.   C.
ANSWER = C

20.

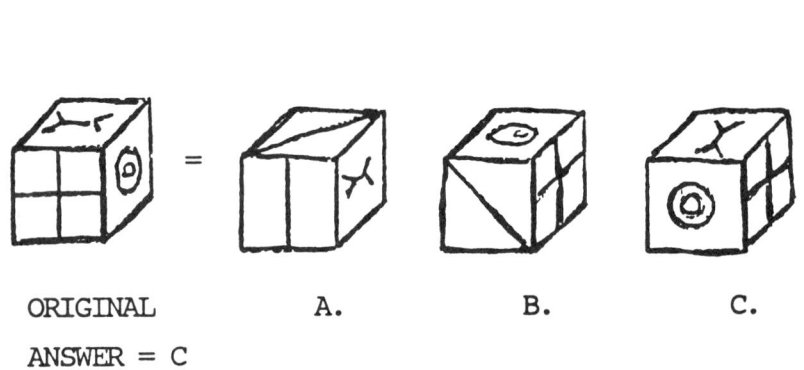

ORIGINAL   A.   B.   C.
ANSWER = C

21.

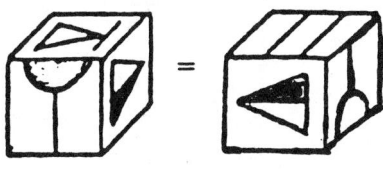

ORIGINAL   A.   B.   C.

ANSWER = A

22.

ORIGINAL   A.   B.   C.

ANSWER = C

23.

ORIGINAL   A.   B.   C.

ANSWER = A

24.

ORIGINAL   A.   B.   C.

ANSWER = B

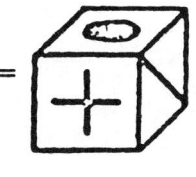

25.

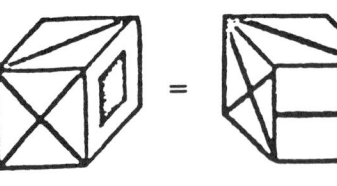

ORIGINAL   A.   B.   C.

ANSWER = C

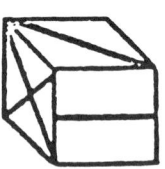

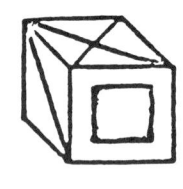

MATCHING FORMS AND FIGURES

IN THE FOLLOWING DRAWINGS DETERMINE THE RELATIONSHIP BETWEEN THE FIGURES LOCATED IN THE TOP TWO BOXES AT THE LEFT, THEN DETERMINE HOW THE FIGURE IN THE BOTTOM LEFT BOX IS SOMEHOW RELATED TO ONE OF THE FOUR FIGURES TO THE RIGHT: (REMEMBER POSITION & DIRECTION)

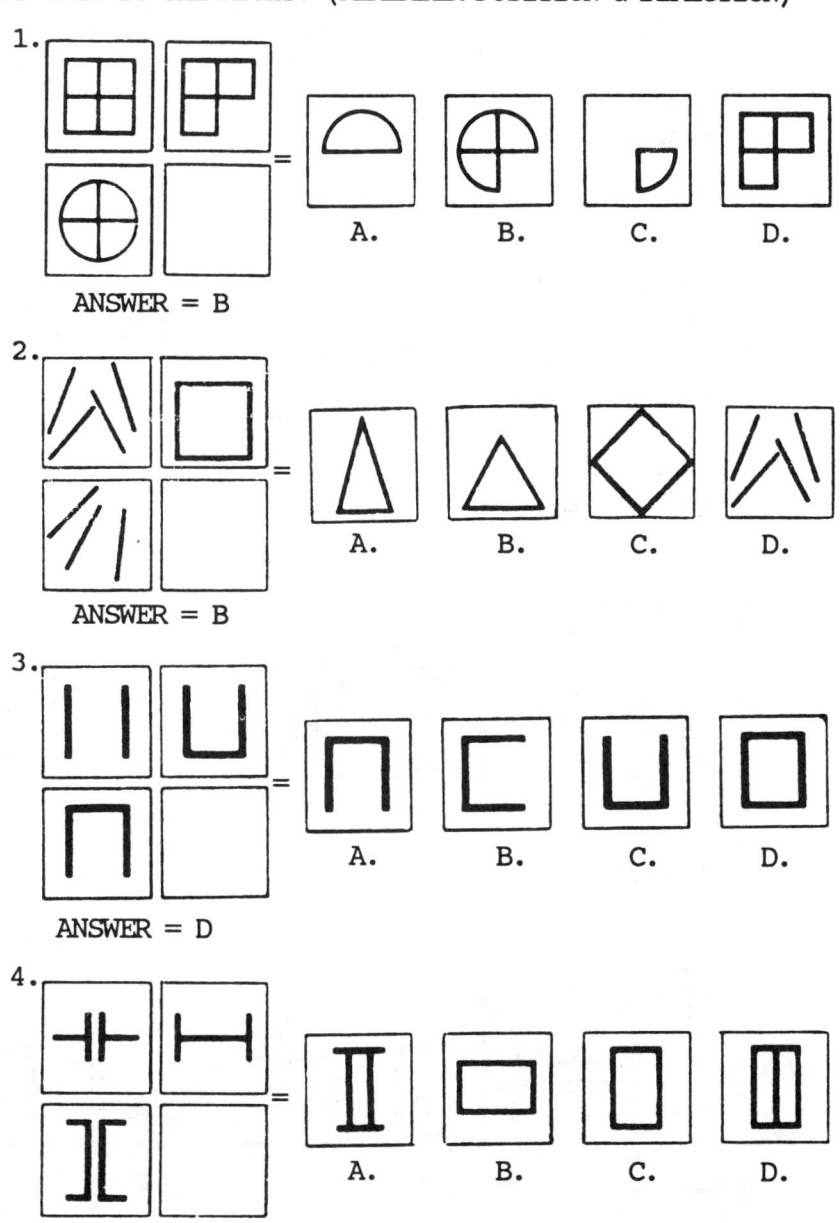

1. ANSWER = B

2. ANSWER = B

3. ANSWER = D

4. ANSWER = C

237

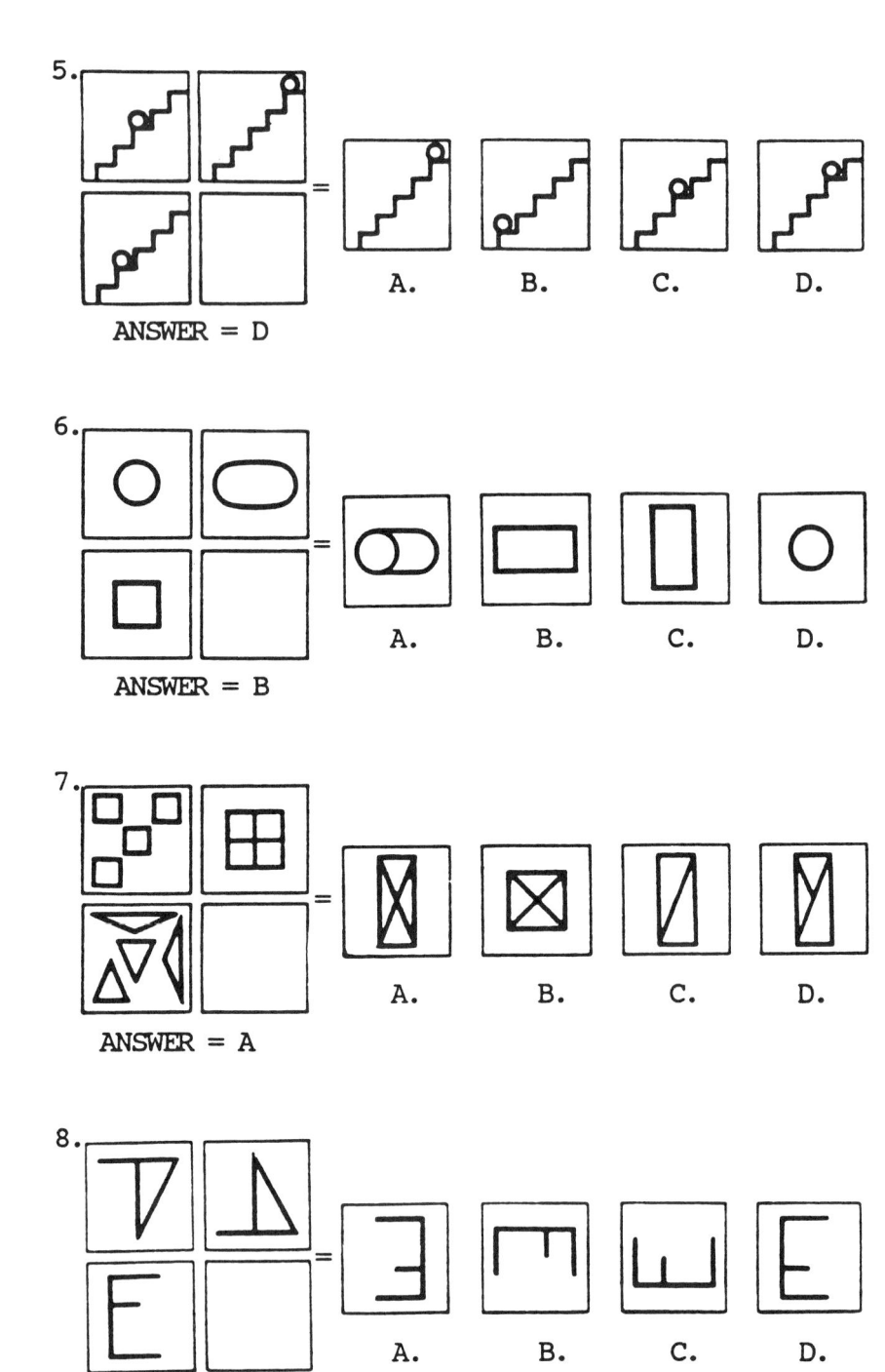

5.

ANSWER = D

6.

ANSWER = B

7.

ANSWER = A

8.

ANSWER = D

9.

ANSWER = A

10.

ANSWER = D

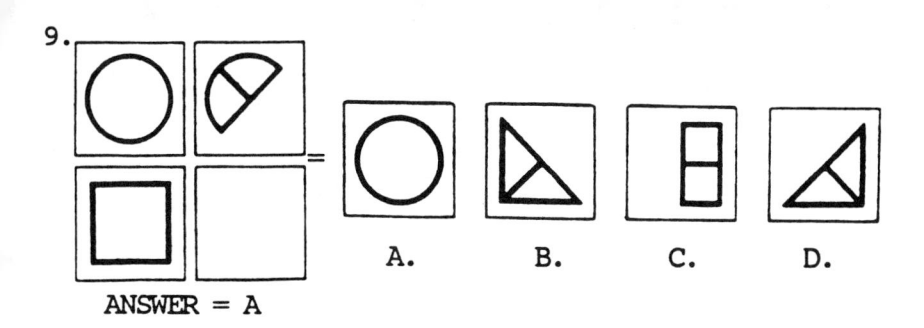

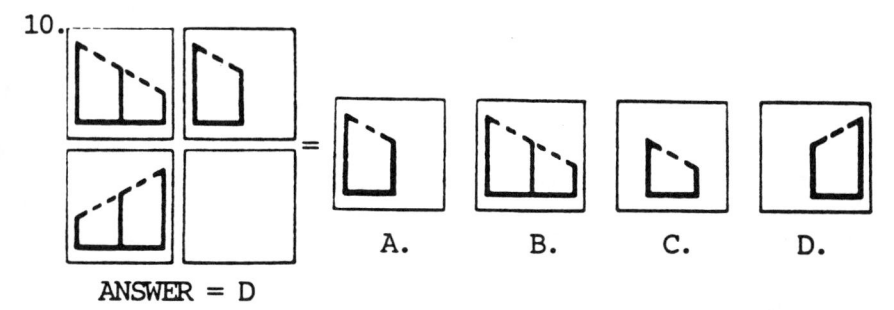

IN THE FOLLOWING DRAWINGS, CHOOSE AN ARTICLE FROM THE RIGHT THAT MAY BE MADE FROM THE FLAT DESIGN AT THE LEFT:

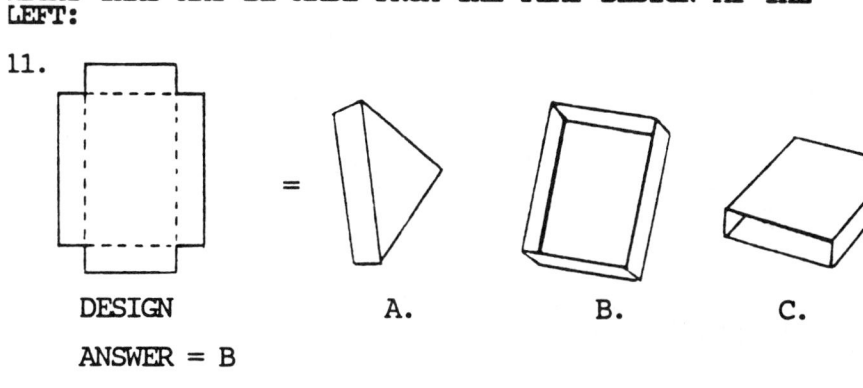

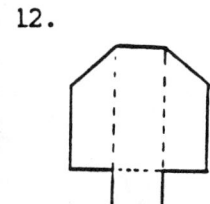

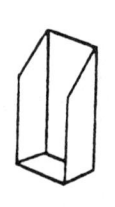

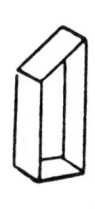

11.

DESIGN A. B. C.

ANSWER = B

12.

DESIGN A. B. C.

ANSWER = B

13.

DESIGN A. B. C.

ANSWER = A

14.

DESIGN A. B. C.

ANSWER = B

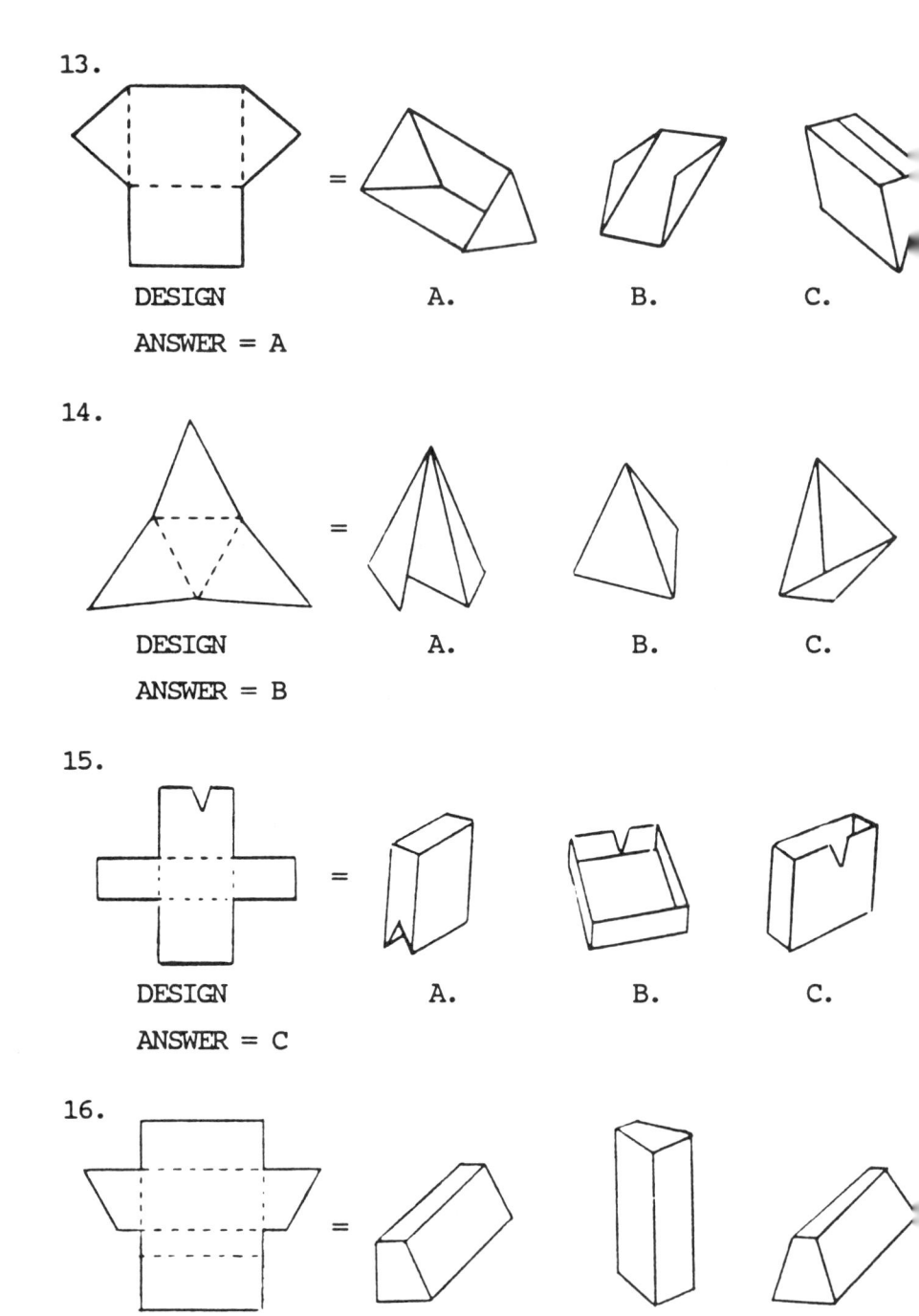

15.

DESIGN A. B. C.

ANSWER = C

16.

DESIGN A. B. C.

ANSWER = A

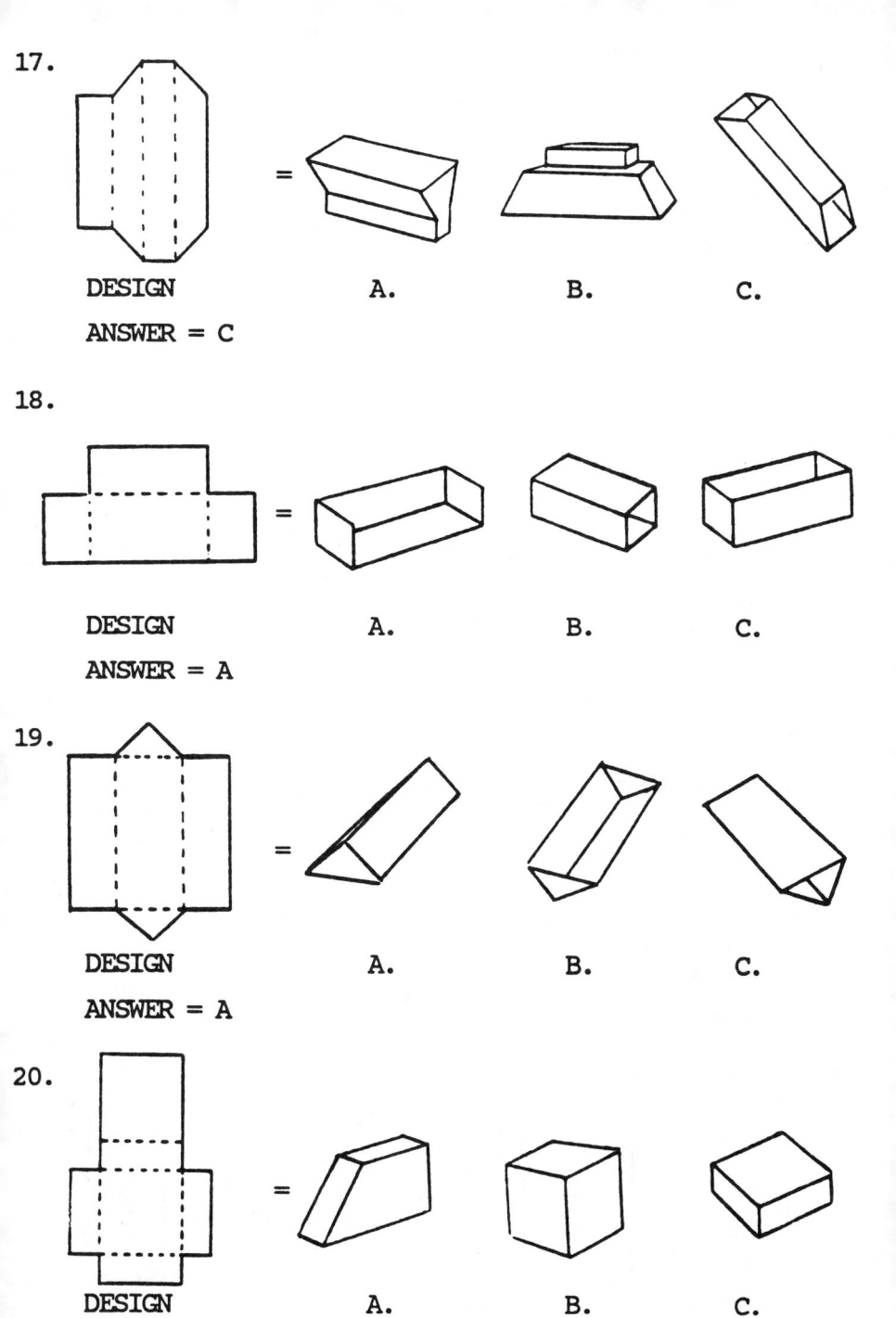

17.

DESIGN A. B. C.

ANSWER = C

18.

DESIGN A. B. C.

ANSWER = A

19.

DESIGN A. B. C.

ANSWER = A

20.

DESIGN A. B. C.

ANSWER = C

21.

DESIGN   A.   B.   C.

ANSWER = B

22.

DESIGN   A.   B.   C.

ANSWER = C

23.

DESIGN   A.   B.   C.

ANSWER = A

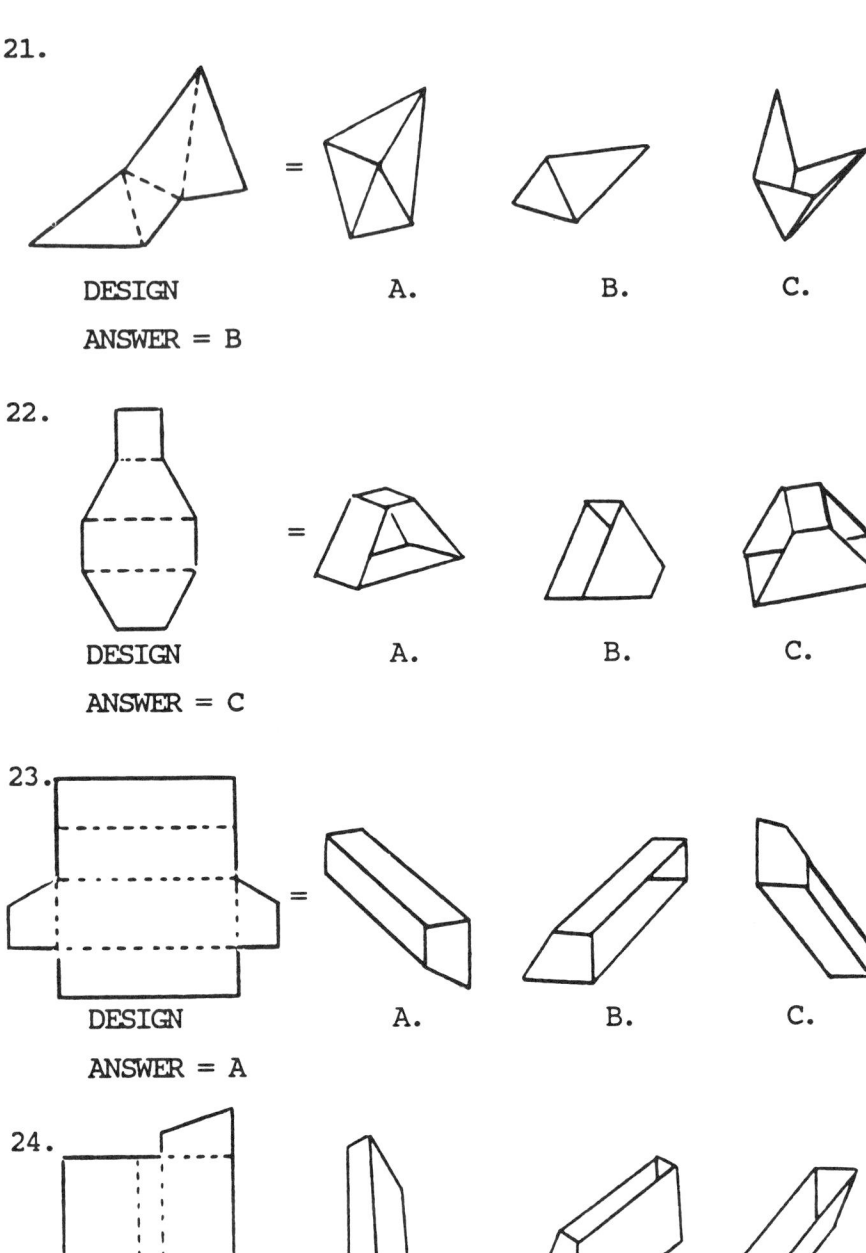

24.

DESIGN   A.   B.   C.

ANSWER = C

25.

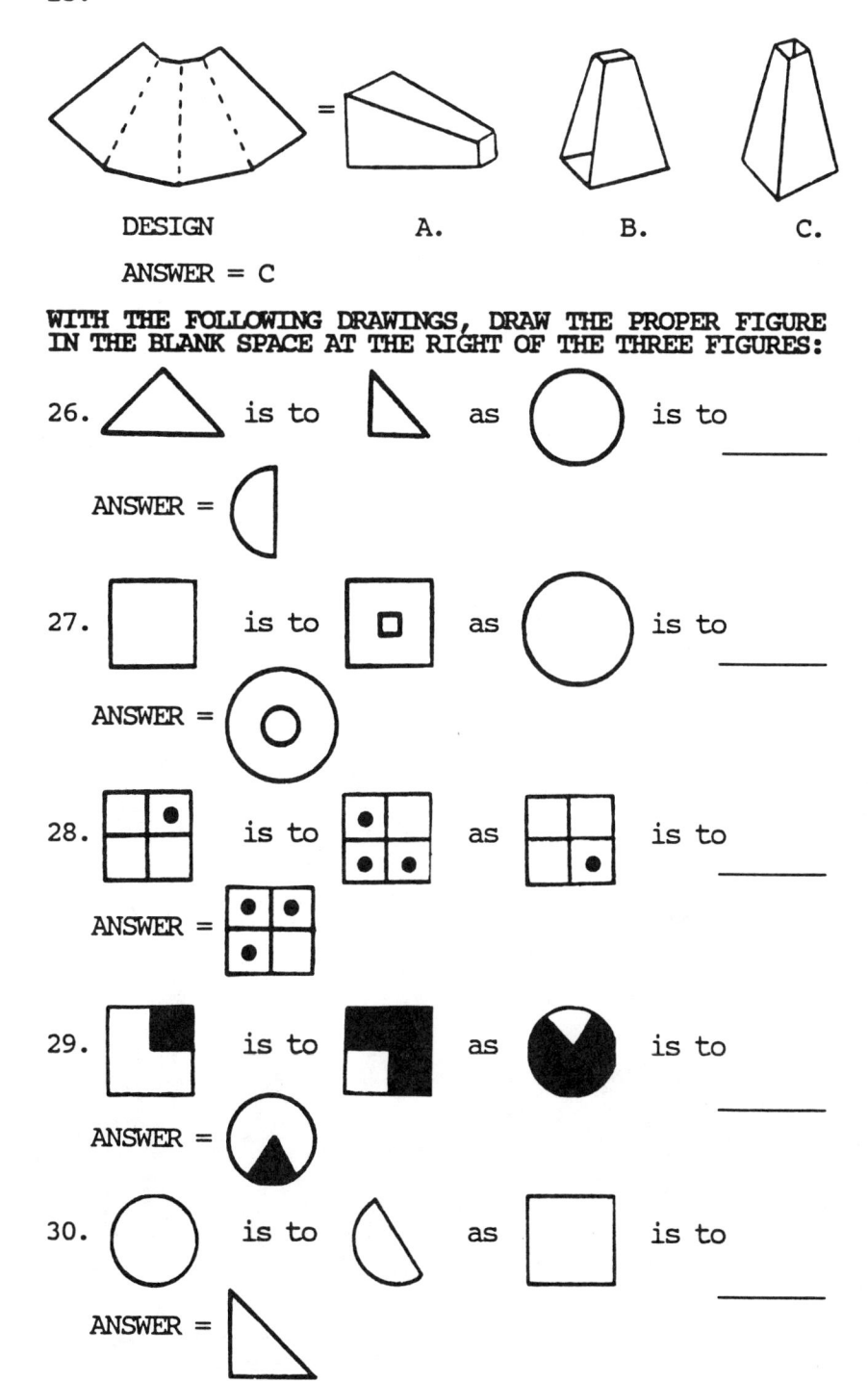

DESIGN   A.   B.   C.

ANSWER = C

WITH THE FOLLOWING DRAWINGS, DRAW THE PROPER FIGURE IN THE BLANK SPACE AT THE RIGHT OF THE THREE FIGURES:

26. △ is to ◁ as ◯ is to ____

   ANSWER = ◐

27. ☐ is to ▣ as ◯ is to ____

   ANSWER = ⊙

28. (square with dot in upper right) is to (square with three dots) as (square with one dot bottom right) is to ____

   ANSWER = (square with three dots)

29. (square with black L-shape) is to (inverted: square with white corner) as (circle with black pie, white wedge) is to ____

   ANSWER = (circle with white, black wedge)

30. ◯ is to (half circle) as ☐ is to ____

   ANSWER = (right triangle)

243

SECTION 11

FIRE SERVICE KNOWLEDGE

EXPLANATION

This part of the entrance level firefighters written exam will cover the general purpose of fire departments and firefighting. Firefighters are responsible for the extinguishment, control, and prevention of fires. Therefore expect questions that relate to these subjects along with other related subjects and duties.

When encountering the portion of the exam that covers fire service information and general knowledge: read each question and all the answer choices carefully prior to deciding on the appropriate answer. Select this answer on the basis of the individual situations described in the question and in the light of universal fire department theory.

Use common sense when dealing with foreign areas. Remember that what seems to be the most logical answer will typically be the appropriate answer. If you have problems with a question, go on to the next question and come back to the difficult questions later if you have time.

If you were not given a handout of fire department policy and procedures for firefighting and other particular knowledge pertaining fire etc., don't panic because you do not have to have a detailed knowledge of the fire departments regulations and firefighting procedures. With a general understanding of basic fire department principles and with good judgement, you should be able to work out the correct answers.

DRIVING

REACTION distance is the distance a vehicle travels between the time the driver sees the hazard and the point where he depresses the brake pedal.

BRAKING distance is the first retardation of the brake to the point where the vehicle stops.

REACTION time = 3/4 of second.

PERCEPTION time = 3/4 of second.

PERCEPTION time + REACTION time = 1 1/2 second.

Total STOPPING distance = perception + reaction + braking distance.

SLOW DOWN while approaching curves.

Emergency apparatus should travel in the LEFT LANE on two lane roadways.

Follow ONE vehicle length per 10 MPH.

A driver can tell the general meaning of a traffic sign by its SHAPE.

STANDARD SHAPE OF SOME ROAD SIGNS:
    1. Octagon   = Stop
    2. Round   = Railroad
    3. Rectangle = Information
    4. Square   = No left turn
    5. Diamond   = Curve or narrow bridge

When approaching a DIAMOND shaped road sign: slow down and drive with caution.

A FLASHING RED signal means to stop and proceed when clear.

When an apparatus goes from a dry part of the road to a SLIPPERY part, let the apparatus continue in gear but stop feeding the fuel.

For each 10 MILES of your apparatus speed you should allow ONE vehicle length.

Most hazardous time to drive is at DUSK.

Centrifugal force can cause a vehicle to go in a STRAIGHT LINE.

When responding CODE 3, minimum following distance = 200 FEET.

Headlights should be ON while responding.

Anticipate hazards created by other drivers in order to comply to the concept of DEFENSIVE DRIVING.

While driving fire apparatus under emergency conditions you may assume that the drivers of other vehicles on the road will behave in an UNPREDICTABLE manner.

A background of TRAINING will best develop good judgement in driving.

While driving fire apparatus you should consider every city intersection as a SEPARATE PROBLEM.

While driving fire apparatus, you should be aware that the term RIGHT OF WAY is the PRIVILEGE of the immediate use of the roadway.

While driving fire apparatus through intersection, you should obtain your LOWEST SPEED at the NEAREST CROSSWALK.

Fire apparatus should be in the NEAREST LANE to the direction it is going to be turned, when approaching an intersection that you plan to make a turn.

Signal for turns at least 100 FEET before the turn and continuously.

The BLIND SPOT of fire apparatus is at the region to the rear left of the apparatus.

While driving a fire apparatus, SAFE SPEED is determined by the EXISTING CONDITIONS.

If a fire apparatus must hit a bump, it is best for the front wheels to hit the bump at a SLIGHT ANGLE.

The STOPPING DISTANCE of fire apparatus is ordinarily affected mostly by speed.

Cross railroad tracks at a SLIGHT ANGLE.

A fire department pumper SHOULD be able to reach a speed of not less than 50 MPH.

The OPTIMUM SPEED for emergency responses by fire apparatus in city traffic is 30 MPH.

When starting a diesel powered apparatus in COLD WEATHER, you should let the engine idle for five to ten minutes.

When starting a diesel engine in WARM WEATHER, you should idle the engine for three to five minutes.

When starting a DIESEL engine, you should crank the engine for thirty seconds at a time and allow two minutes between cranking attempts.

LUGGING, is when an engine is operated at full throttle (accelerator fully depressed) below rated speed.

Operating an engine COLD can cause:
    1. High piston wear.
    2. Increased oil consumption.
    3. Increased valve deposits.
    4. Shorter engine life.

The practice of starting fire engines at change of shift is unnecessary and of no practical value with modern engines. REASONS:
    1. Causes needless wear and abuse.
    2. Does not prove engine will start the next time.
    3. Large portion of engine wear takes place during first few seconds of starting.
    4. When engine is cold, water and fuel condense on working parts, causing corrosion of valve lifters, springs, and other internal parts.
    5. Water vapor mixes with carbon and dirt particles in oil forming emulsion to sludge.

It is NOT advisable to "GUN" a gasoline engine before shutting down primarily because it will cause crankcase dilution with the gasoline.

LADDERS

EXTENSION LADDERS carried on **LADDER TRUCKS**:
1. One 14 foot
2. One 28 foot.
3. One 35 foot.
4. One 40 foot.

STRAIGHT LADDERS carried on **LADDER TRUCKS**:
1. One 16 foot.
2. One 20 foot.

LADDER COMPANY: a fire company that operates a ladder truck and is trained in ladder work, ventilation, rescue, forcible entry, and salvage work.

N.F.P.A 193 recommendations for ladder capacities:
LADDER SIZE   # MEN
1. Folding ladder (attic ladder)   = 1
2. Small extension ladders (inside use) = 1
3. 16 feet to 26 feet extension ladders = 2
4. 26 feet to 35 feet extension ladders = 4
5. 36 feet to 45 feet extension ladders = 5
6. 45 feet and above extension ladders = 6

Common sizes of **AERIAL LADDER TRUCKS** are:
1. 65 feet.
2. 75 feet. (minimum size recommend for fire service).
3. 85 feet.
4. 100 feet.

AERIAL LADDERS minimum desirable size for fire use is 75 feet.

TRUSS means to support, strengthen, or stiffen as a beam.

Ground ladders on **LADDER TRUCKS** should equal a minimum of 163 feet, total.

GROUND and **AERIAL LADDERS** should be raised three or more rungs above the roofs edge.

Distance ladder should be from building formula = the used length of ladder divided be four.

250

Ladder locks, pawls, or dogs when **ENGAGED**:
1. Take the load off the cables.
2. Line-up the ladder rungs.

LADDER TIP in windows should be at either side and above the sill.

AERIAL TRUCK: a ladder truck with a permanently mounted aerial ladder in addition to its regular equipment.

ARTICULATING boom platforms have two or more sections called **"BOOMS"**.

"BOOMS" are either tubular truss-beam or steel box-beam construction.

Aerial hydraulic **system oil** performs 3 functions:
1. Transmission of power.
2. Lubrication of all parts.
3. Cooling of operating pump and motors.

HYDRAULIC FLUID: non-compressible liquid which transfers pressures from one point to another.

Horn or buzzer signals from tiller operator or driver on **LADDER TRUCK:**
1. One blast   =   Stop.
2. Two blast   =   Proceed forward.
3. Three blast =   Reverse.

AERIAL LADDERS best climbing angles are between 70 degrees and 80 degrees.

CANTILEVER POSITION is when the upper end is unsupported.

The hoisting cylinders on **AERIAL APPARATUS** are sometimes referred to as elevating cylinders.

AERIAL LADDER spotted in jackknife position of 60 degrees from incline position will provide excellent stability, without blocking most streets, and will provide:
1. Close in spots.
2. Good climbing angle.
3. Maximum ladder strength.
4. Greatest ladder reach.

TRACTOR TRAILER type Aerial Ladder is more:
1. Maneuverable.
2. Stable.
3. Space for equipment.
4. Positioned faster.

Fire Department ladders should be made of **METAL**.(usually aluminum).

FIRE DEPARTMENT LADDERS SHOULD BE TESTED:
1. At regular intervals
2. After each use.
3. After major repairs.
4. At least once a year.

Test load on top quality ladder should be twice that permitted for ordinary use.

FOLDING LADDERS are most often used to gain access to attics or that portion of a building above the ceiling.

ATTIC LADDER = 8 feet to 14 feet long.

ATTIC LADDER: small ladder that can be folded to less than a foot for storage and passage in cramped quarters.(also known as folding ladder).

BUTT of ladder = bottom or ground end.

HEEL of ladder = bottom or ground end. (also).

BEAMS of ladder = the two principle structural sides.

RUNGS of ladder = the cross members between the beams on which people climb. (usually round).

HOOKS of ladder = curved, sharp, metal devices that will fold outward from each beam at the top end of roof ladders.

BED of ladder = the lower section of an extension ladder.

FLY of ladder = the upper or top sections of an extension ladder.

TRUSSED type ladders have greater strength and rigidity than solid side ladders.

FIRE PUMPS

The **COMPOUND GAGES** which are installed on the suction side of fire pumps are calibrated to measure vacuum in inches of mercury and pressure in pounds per square inch. (PSI)

BAROMETRIC PRESSURE will affect the drafting ability of a pump.

The main **FEATURE** of **RELIEF VALVES** are their **SENSITIVITY** to change in pressure and their ability to relieve this pressure within the pump discharge.

The main **PURPOSE** of the **RELIEF VALVE** which is installed on a pumper engine is to permit water to flow from the discharge to the suction side of the pump.

The main **FEATURE** of a **GOVERNOR** is that it regulates power output of the engine so that it matches pump discharge requirements.

The type of gage most commonly used on the suction side of centrifugal fire pump is the **DOUBLE-SPRING COMPOUND GAGE.**

The capacity of a **POSITIVE-CAPACITY** pump is limited by its displacement and revolutions per minute.

The **POSITIVE DISPLACEMENT** pump is seldom used to produce fire streams in modern times.

For producing fire streams, the **POSITIVE DISPLACEMENT** pump has largely been replaced by the **CENTRIFUGAL** pump.

The most simple **POSITIVE DISPLACEMENT** pump is the **PISTON** pump.

A pump that has a piston with a **BACKWARDS** and **FORWARD** motion is called a **RECIPROCATING PUMP.**

A **POSITIVE DISPLACEMENT PUMP** will have a measurable capacity per stroke revolution.

CENTRIFUGAL means proceeding away from the center, developing outward, or impelling an object outward from a center of rotation.

In a **CENTRIFUGAL** pump the rotating wheel is known as the **IMPELLER**.

Each impeller and housing in a **CENTRIFUGAL** pump is called a **"STAGE"**.

As used in the fire department pumping manuals, the term **"NET PUMP PRESSURE"** is best defined as the pressure actually produced by the pump.

1500 GPM is the **HIGHEST STANDARD** size GPM pumper.

TRIPLE COMBINATION PUMPER: hose, water, and pump.

QUAD: hose, water, pump, and ground ladders.

QUINT: hose, water, pump, ground ladders, and aerial ladder.

GAWR: gross axle weight rating.

GCWR: gross combination weight rating.

GVWR: gross vehicle weight rating.

Pumps rated capacity is determined by **TESTING**.

The capacity of a **POSITIVE DISPLACEMENT** pump is limited by its displacement.

If discharge valve is closed on **CENTRIFUGAL** pump the load is decreased on the motor.

VACUUM: space completely void of matter.

HEAT EXCHANGER = "Indirect auxiliary cooler system".

RADIATOR COOLER = "Direct auxiliary cooler system"

CENTRIFUGAL PUMPS are not able to create a vacuum, therefore they need **PRIMING DEVICES**.

CENTRIFUGAL PUMP PRINCIPLE: rapidly revolving disc will tend to throw a liquid from the center toward the outer edge of a disc.

CENTRIFUGAL PUMPS are pumps with one or more impellers that rotate on a shaft, taking water into the eye of the impeller and discharges through the volutes. Centrifugal pumps may be single stage or multiple stage.

In **CENTRIFUGAL PUMPS**: with the quantity remaining constant, the pressure will increase at a rate equal to the square of the speed increase.

In both positive and single stage centrifugal pumps: with the pressure remaining constant, the quantity of discharge is **DIRECTLY PROPORTIONAL** to the speed of the pump.

CENTRIFUGAL PUMP PRINCIPLE: Tendency of revolving body to fly outward from the center of rotation.

CENTRIFUGAL PUMP: water is expelled from one place on the perimeter.

POSITIVE DISPLACEMENT PUMP PRINCIPLE: Incompressibility of water (1% to 30,000 lbs pressure)

CENTRIFUGAL PUMPS: Employ a certain principle of force in pumping. The power or force to create pressure is exerted from the center. The revolving motion of the impeller will whirl water introduced at the center toward the outer edge of the impeller. Here it is trapped by the pump casing and is forced to the discharge outlet.

In **CENTRIFUGAL PUMPS**, power is transmitted from the drive shaft, through the pump transmission, and intermediate gear to the impeller shaft.

VOLUTE handles the increasing volumes of water.

VOLUTE is the design of water passageway in a centrifugal pump.

MAIN ACTION of any pump is to add pressure to the water.

VOLUTES primary purpose is to handle an increasing volume of water.

In **POSITIVE DISPLACEMENT PUMP** the pump speed is less than the engine speed. (greater efficiency)

In a **CENTRIFUGAL PUMP**, the pump speed is greater than then engine speed.

One of the major **DIFFERENCES** between positive displacement pumps and centrifugal pumps in fire fighting apparatus is that centrifugal pumps will not pump air.

With **BOTH** centrifugal and positive displacement pumps, the quantity of water which is discharged is directly proportional to the pump speed when the pressure is constant.

HEAD: pressure due to elevation.

Changing pressure on a single stage centrifugal pump depends on the **MOTOR SPEED**.

The **MAIN ADVANTAGE** that a centrifugal pump has over positive displacement pumps is that the centrifugal pump can exceed its rated capacity.

CENTRIFUGAL PUMPS have the fewest parts to wear out of any fire service type pump.

In a centrifugal pump the quantity of water issuing from the impeller remains **CONSTANT** throughout the entire rotation.

CENTRIFUGAL PUMP: cannot pull a vacuum because it works on the principle of **NON-DISPLACEMENT**.

CENTRIFUGAL PUMPS: are less likely to be damaged by fire pump operators because they only churn water in the within the pump chamber when hose lines are shut down.

In a centrifugal pump, the energy being imparted to the water initially creates **VELOCITY**.

In positive displacement pumps **SLIPPAGE** is dependent on the condition of the pump and the operating pressure.

PARALLEL OPERATION: each impeller discharges into common outlet = increased flow, reduced pressure.
When the transfer valve of a series-parallel pump is placed in the **PARALLEL** position, water from the first stage impeller is routed directly to the pump discharge.

After water has passed through the impeller into the pump volute, it is mainly prevented from returning to the suction side of the pump by the **VELOCITY** of the water.

In origin pump damage from cavitation = **MECHANICAL**. In a series-parallel centrifugal pump the **CLAPPER** check valves are closed by the first stage pressure.

The **TRANSFER VALVE** on fire apparatus is used to allow the pump to operate most effectively.

SERIES-PARALLEL PUMPS are commonly utilized in fire apparatus because they are capable of providing **GREATER VARIATIONS** in capacity and pressure.

Two impellers working in **SERIES** in a two stage pump will give reduced volume at a higher pressure.

A multi-stage centrifugal pump uses two or more impellers to build-up **PRESSURE**.

When pumping from a hydrant supply, the rated capacity of a centrifugal pump is normally **EXCEEDED**.

Compared with positive displacement pumps, centrifugal pumps are generally **LESS EFFICIENT**.

Compared with centrifugal pumps, **ROTARY GEAR PUMPS** generally are more suitable for pumping air.

The object of **PACKING** on a pump is to prevent an air or liquid leak.

The capacity rating given to a pumper is on the basis of delivering full rated capacity from draft at a **10 FOOT LIFT**.

Pump **CAVITATION** occurs mainly in the **IMPELLER EYE**.

A **CLAPPER VALVE** is an automatic valve installed in hydraulic systems, which permits the flow of liquid in **ONE DIRECTION ONLY**.

Only wearing parts of centrifugal pumps are the **BEARINGS**.

VACUUM PUMP: removes air or other gases.

FLUCTUATING PRESSURE is called **"HUNTING"**.

CAVITATION can occur whenever a pump is used improperly for existing conditions:
1. Delivering more water than can enter the pump.
2. Excessive lift.
3. Suction hose too small for amount of discharge.
4. Suction blocked at strainer.
5. Suction collapse.
6. Water temperature too high. (above 85 degrees F)
7. Low atmospheric pressure. (high altitudes)
8. A combination of any of one or all of these conditions, at any tank, hydrant, relay, or drafting operation.

PULSATING PRESSURE: indicates air leaks, or restricted suction.

If discharge valves are **CLOSED** on centrifugal pumps, the load is **DECREASED** on the motor

RATED CAPACITIES OF PUMPERS:
1. 2000 GPM.
2. 1750 GPM.
3. 1500 GPM.
4. 1250 GPM.
5. 1000 GPM.
6. 750 GPM.
7. 500 GPM.

When pumping it is safer to stand to the **INSIDE** of the bend in charged hose line.

Pumpers **SHOULD** have service test annually and after major repairs.

Pumps that are rated **LESS THAN 500 GPM** and that are permanently mounted are called **BOOSTER PUMPS.**

The **SMALLEST** size pump recognized for a fire pumper is 500 GPM.

ROTARY VANE hydraulic motors operate both clockwise and counter clockwise.

CERTIFICATION TEST AT DRAFT:
1. 100% volume @ 150 PSI for 2 hours.
2. 100% volume @ 165 PSI for 10 minutes. spurt
3. 70% volume @ 200 PSI for 30 minutes.
4. 50% volume @ 250 PSI for 30 minutes.

DELIVERY TEST (ACCEPTANCE) AT DRAFT:
Same as certification test, plus driving performance, carrying capacity, cooling system, suspension, and braking system. This is the road test. Test should be conducted by joint supervision of representatives from the manufacture and the Fire Department. This is considered the **MOST IMPORTANT TEST**, it is the baseline for later comparisons with the service test.

SERVICE TEST AT DRAFT:
1. 100% volume @ 150 PSI for 20 minutes.
2. 100% volume @ 165 PSI for 5 minutes. (not required)
3. 70% volume @ 200 PSI for 10 minutes.
4. 50% volume @ 250 PSI for 10 minutes.

A **MULTI-STAGE** centrifugal pump is equipped with **TWO** or **MORE** impellers.

A two stage of centrifugal pump would most likely have two impellers connected in **SERIES**.

Engine pumper must be able to produce its rated capacity at **80%** or less of its **PEAK** engine speed and must be able to produce its rated pressure at **90%** or less of its **PEAK** engine speed.

The **ACTUAL CAPACITY** of a centrifugal pump is limited by its design: intake diameter, impeller, eye of impeller, diameter of impeller eye, width of impeller, shape and number of vanes in the impeller, and the design of the volute chamber.

TRANSFER VALVE changes pump from volume to pressure, and vice-versa.

SHROUD is the casing of the impeller.
Impellers are made of **BRONZE**. Impeller shafts are made of **STAINLESS STEEL**.

Centrifugal pump **WEAR RINGS** (clearance rings) are positioned nearest the impeller eye.

CENTRIFUGAL PUMP IMPELLER:
1. Large eye = **MORE GPM**
2. Wider impeller = **MORE GPM**
3. More vanes = **MORE GPM**
4. Less vane curvature angle = **MORE GPM**

VOLUTE: enables centrifugal pump to handle the increasing quantity of water towards the discharge outlet and at the same time permit the velocity of the water to remain constant or to decrease gradually maintaining the continuity of flow.

WATER UNDER PRESSURE, in the volute of a centrifugal pump is prevented from returning to the suction side of the pump by close fit of the impeller hub to a stationary clearance (wear) ring at the eye of the impeller, and hydraulic pressure due to the velocity created by the centrifugal force.

VOLUTE = progressively expanding waterway which converts velocity to pressure as the velocity remains constant.

VOLUTE PRINCIPLE is the design of the water passageway in centrifugal pumps.

A centrifugal pump will create its **HIGHEST VACUUM** when the pump is filled with water.

FIRE STREAMS/HOSE

Types of **HOSE STREAMS:** fog and straight; master streams and hand lines.

CHIEF PURPOSE of a nozzle on a fire stream is to give it shape and added velocity.

The **PERFECT** fire stream cannot be sharply defined.

The time that a stream of water passes through the air it is influenced by, its **VELOCITY**, and by: **WIND, GRAVITY,** and **FRICTION WITH THE AIR.**

BROKEN FIRE STREAM: stream of water that has been broken into coarsely divided drops.

MASTER FIRE STREAM: any fire stream that is too large to be controlled without the use of mechanical help. **MOBILITY** of a fire stream determines its effectiveness.

MASTER STREAM DEVICES mounted on the deck of a fire apparatus and directly connected to the pump:
1. Turret pipe.
2. Deck gun.
3. Deck pipe.

MONITOR: a master stream device with the ability to change the direction of its stream while water is flowing.

PENETRATION AND DEFLECTION: determine the effectiveness of master streams.

FOG STREAMS have a larger diameter than solid streams, therefore there is more area that has to deal with air friction loss, thus causing its forward velocity to be retarded more rapidly.

FLOW PRESSURE: is the rate of flow of the water from a discharge opening.

Maximum stream **PENETRATION ANGLE** is at **45 DEGREES.** If the angle increases the penetration decreases.

Greatest **HORIZONTAL REACH** occurs at **30 - 34 DEGREES** angle.

Maximum effective **VERTICAL REACH** occurs at **60 - 75 DEGREES** angle.

MAXIMUM VERTICAL REACH is attained when the nozzle is perpendicular to the ground, but is not used, maximum angle of common vertical stream is considered to be **60 - 70 DEGREE** angle.

HIGHEST floor to direct fire stream from ground level is to the **THIRD FLOOR**.

ANY wind will hinder a fire stream.

An **EFFECTIVE FIRE STREAM** discharges **90%** of its volume inside a circle **15"** diameter, and **75%** of its volume inside a circle **10"** in diameter.

Hand line **MAXIMUM** penetration distance = 50' into structure.

Excessive pressure will break-up a small stream **FASTER** than a large stream.

Effectiveness of **MASTER STREAM** is the penetration and deflection.

EFFECTIVENESS of any fire stream = **MOBILITY** of the stream.

FIRE STREAM: a stream of water from the time it leaves a nozzle until it reaches the point of intended use.

Course of a fire stream is **AFFECTED** by:
1. Gravity.
2. Friction due to air resistance.
3. Wind velocity.
4. Obstacles.

With adequate reach, the effectiveness of a fire stream, as far as fire extinguishing is concerned, increases the most as the stream takes on the characteristic of a **FOG STREAM**.

The reaction caused by a hose nozzle is due to **VELOCITY**.

Spray for fire streams in **CONTRAST** to solid fire streams:
  POSITIVE POINTS:
  1. Absorbs more heat more rapidly.
  2. Covers greater area with water.
  3. Uses less water.
  NEGATIVE POINTS:
  1. Requires higher discharge pressure.
  2. Has shorter reach.
  3. Less penetration.
  4. Less cooling effect in subsurface areas. (charred wood, etc.)

Nozzle reaction on **SPRAY NOZZLES** is less than a straight stream because the smaller impinging streams will defuse the reaction forces.

DEFECTIVE hose stream caused by:
  1. Too much pressure.
  2. Too little pressure.
  3. Air in line.
  4. Kinks in hose.
  5. Hose twisted near nozzle.
  6. Defective nozzle.

COMMON DESIGNS OF NOZZLE CONTROL VALVES:
  1. Rotary.
  2. Ball valves.

SOLID STREAM: is a stream that stays together in a solid mass.

SPRAY STREAM: finely divided particles of water.

STRAIGHT STREAM: a solid stream of water used to gain maximum water force against an object or to maximize the water reach and penetration.

FOG STREAM: a billow of water mist discharged by a spray nozzle.

NOZZLE PRESSURE: the velocity pressure in PSI at which water is discharged from a nozzle.

WYE CONNECTION: is used to divide one hose line into two or more hose lines.

The **WYE** has two male branches from a female stem.

SIAMESE FITTING: is employed to bring two or more hose lines into a single hose line.

The **SIAMESE** fitting has a male stem and two female branches.

The **SIAMESE** fitting when used, should be as close to the fire as possible.

FITTINGS = hardware used to connect hose lines:
1. Reducers.
2. Increasers.
3. Adapters.
4. Couplings.

Both **INCREASERS** and **STRAIGHT REDUCERS** have male and female ends of a different size.

The **FEMALE** end is the smaller end on **INCREASERS** but not on reducers.

A **REDUCING WYE FITTING** has one male connection and two female connections of a **SMALLER** size.

When hose bursts under pressure it can start to whip around; the best place for a firefighter to be is **INSIDE** the bend.

FIRE HOSE MUST:
1. Withstand high pressures.
2. Transport water with a minimum loss in the working pressure.
3. Be flexible enough so that it can be handled and used under extreme fire conditions without requiring a large number of firefighters.

Store hose loosely rolled in a cool dry place.

If you must drive over fire hose, it is best to use a **HOSE BRIDGE**.

If you must drive over fire hose **WITHOUT** the use of a hose bridge, the hose should be charged with water under pressure. The most damage occurs when the fire hose is empty.

As a general rule, the diameter of a plain nozzle to be used on a line should **NOT EXCEED** 1/2 of the diameter of the hose.

Smallest tip for **MASTER STREAM** = 1 1/4".

DOUGHNUT ROLL: 50' length of hose doubled and rolled toward coupling.

MINIMUM amount of reserve hose = one full load.

Amount of hose needed for **FIRE BUILDING** = 1 length per story plus one length.

Each **ENGINE COMPANY** should carry:
1. 1200 feet of 2 1/2 inch hose or longer.
2. 400 feet of 1 1/2 inch hose.
3. 200 feet of 1 inch hose.

DOUBLE JACKET hose will permit minimum twisting.

DOUBLE JACKET hose's greatest advantage over single jacket hose is durability.

Main reason for double jackets on fire hose is to allow **HIGHER** working pressures.

The ability of fire hose to withstand the high pressures they undergo, without bursting, is because of the **COTTON JACKET**.

COTTON JACKET hose is the most susceptible hose to mildew.

Advantage of **MULTIPLE JACKET HOSE** over single jacket hose is that it has the ability to withstand a greater amount of chafing.

DACRON FILLER HOSE has greater friction loss than all cotton jacket fire hose.

The main advantage that **DACRON** fire hose has over cotton jacket hose is that it is **LIGHTER** in weight.

Cotton double jacket hose should be **TESTED ANNUALLY**
1. At 250 PSI.
2. For 5 minutes.
3. In 300 foot lengths.

Fire hose should last a **MINIMUM** of **10 YEARS**.

HARD SUCTION should withstand a working pressure of **200 PSI**.

Poor hose layout = **LOW CAPACITY**.

N.F.P.A. #196 is the standard for fire hose.

N.F.P.A. #198 is the standard for the **CARE** of fire hose.

Upon receipt of **NEW** fire hose it should be tested at a pressure that complies with **NFPA** standard **196**.

SERVICE testing of fire hose should comply with **NFPA** standard **198**.

Fire hose **SERVICE** test is based on a minimum pressure of **250 PSI**.

The maximum amount of time fire hose should go without a service test is **ONE YEAR**.

Hose line to be tested should not exceed **300 FEET**.

Specifications for fire apparatus **HOSE COMPARTMENTS** are found in **N.F.P.A. 1901**.

Hose compartments on fire apparatus are usually called **HOSE BEDS**.

The divider or separator in a fire hose compartment is called **BAFFLE-BOARD**.

THREE BASIC CONSTRUCTION METHODS OF FIRE HOSE:
    1. Braided.
    2. Wrapped.
    3. Woven.

UNLINED FIRE HOSE is usually used inside buildings on standpipe risers.

The **FEMALE COUPLING** should be connected to the pump discharge for pre-connected hoselines.

2 1/2" hose moves 1 ton of water per minute.

A line of fire hose from which water is flowing through a nozzle, an open butt, or a broken line, and which is not under control by the firefighters is called **A WILD LINE**.

The safest way to control a **WILD LINE** is by **CLOSING A VALVE** to shut off the flow of water.

When advancing hose lines up a ladder, the firefighters should have about **10 FEET** between them.

Where fire hose is being advanced up a ladder there should be **20 - 25 FEET** of hose between each firefighter.

Devices that are used with fire hose but water does not pass through are called **HOSE TOOLS**.

HYDROKINETIC: water in motion; such as water flowing through a hose line.

The **FASTER** that water is traveling forward, the **FARTHER** it will reach before being pulled to the ground by gravity.

50 foot section of 2 1/2" cotton rubber lined hose, filled with water weighs **106 LBS**.
(12.75 gallons X 8.35 LBS = 106 LBS)

1 foot section of 2 1/2" cotton rubber lined hose filled with water weighs **3 LBS** to **4 LBS**, counting the water and hose weight.

50 foot section of cotton rubber lined hose without water and not counting couplings weighs:
    1. Single jacket = **35 LBS**.
    2. Double jacket = **50 LBS**.

FORCIBLE ENTRY/SALVAGE/OVERHAUL/VENTILATION

Regardless of the class of a door, firefighters should be sure that the door is LOCKED before attempting to gain entry to a building with the use of force.

BEFORE attempting to force any door the firefighters should:
1. Check to see if door is locked.
2. See if hinge pins can be removed.
3. Have hose lines available.

BREACHING: the opening of masonry walls.

Wooden joist of a wood floor are usually a maximum of **16 INCHES APART**.

FORCIBLE ENTRY: entry into a secured building with a minimum of delay, often by the use of special tools. (forcible entry tools).

CLASSIFICATION OF DOORS FOR FIREFIGHTERS:
1. Swinging.
2. Revolving.
3. Overhead.
4. Sliding.

WOOD SWINGING DOORS:
1. Panel.
2. Slab.
3. Ledge.

FORCIBLE ENTRY TOOLS: tools carried on fire apparatus used to gain entry into buildings and obstructions so that firefighting and rescue operations may be carried out.

The **BEST** way for firefighters to learn to recognize various types of windows is to be involved in through building inspection surveys

The distance from the window sill to the floor is usually about **4 FEET**.

As far as **GLASS-PANELED** doors go, it is **CONSIDERABLY** more expensive to replace **TEMPERED** plate glass than any other of the same size.

Firefighters should use every other available means of forcible entry before trying to gain entry through a **TEMPERED PLATE GLASS DOOR**.

To prevent glass from sliding down an axe handle while a firefighter is breaking a plate glass window, the firefighter should **STRIKE THE UPPER PART OF THE WINDOW FIRST WHILE STANDING TO ONE SIDE**.

To protect contents of a building and reduce water damage during a fire, use **SALVAGE COVERS** over as much of the building and its contents as possible.

CARRYALL: salvage device that is 6 feet or eight feet square with rope handles at the edges.(used to carry or catch debris).

FLOOR RUNNER: usually 3 feet by 18 feet (can be up to 30 feet long). Made of canvas or plastic and used on floors to prevent damage to the floor or carpeting from mud, debris, or water.

When VENTILATING a roof, the firefighter should cut a rectangular shaped hole.

For **FORCED AIR VENTILATION** the opening for the replacement air should be equal to or larger than the opening for the venting hole.

SKYLIGHTS that are made from glass are effective vents because the temperature from the fire will break the glass.

WINDWARD: the side of the building where the wind is hitting.

LEEWARD: the side of the building that is opposite the side where the wind is hitting.

FIRE PREVENTION/BUILDING CONSTRUCTION

U.B.C.: UNIFORM BUILDING CODE is prepared by the International Conference of Building Officials and is published in three year intervals.

U.B.C. : Governs new construction of structures, buildings.

U.F.C. : UNIFORM FIRE CODE is prepared by International Conference od Building Officials and Western Fire Chiefs Association. (technical advisors)

U.F.C. : Governs the maintenance of regulations by Governmental Agencies.

N.B.C. : NATIONAL BUILDING CODE is prepared by A.I.A. which is the American Insurance Association and is published in three year intervals.

N.F.P.A. : NATIONAL FIRE PROTECTION ASSOCIATION is prepared by technically competent committees having balanced representation.

N.F.P.A. : is adopted by public authorities with law making or rule making powers only. (National Fire Codes)

BUILDING CODES are designed to provide rules for public safety in the construction of buildings to the extent which can be applied as law under the broad authority of the police power.

FIRE SAFETY AND CONTROL can best be accomplished by adoption and enforcement of codes and standards.

FIRE LOAD: the expected maximum amount of combustible material in a single fire loss.

STATE GOVERNMENTS do not get involved with forest fire protection.

FEDERAL GOVERNMENT does not get involved with compiling data for insurance and losses.

N.F.P.A. standard calls for automatic smoke vents over theater stages after the "Iroquois" theater fire in 1903.

FIRE PREVENTION: the fire protection activities with the purpose of preventing fire from starting.

FIRE PREVENTION WEEK: a week devoted to publicizing fire prevention activities. During the week of October 9. (date of great Chicago fire)

Public buildings, schools, hospitals, etc. are all types of **TARGET HAZARDS.**

FIRE PREVENTION BUREAU: a unit of a fire department which does its major work in fire prevention and investigation rather than combatting fires.

FIVE TYPES OF BUILDING CONSTRUCTION:
1. Fire resistive.
2. Heavy timber.
3. Non-combustible.
4. Ordinary.
5. Wood frame.

Buildings with **FIRE-RESISTIVE** construction have the greatest resistance to structural damage by fire.
FOUR TYPES OF WALL CONSTRUCTION:
1. Reinforced concrete.
2. Masonry.
3. Steel frame.
4. Wood frame.

FIRE RESISTIVE CONSTRUCTION = structural members including walls, partitions, columns, floor, and roof, are made of non-combustible materials of specific ratings.

FIRE DOORS restrict flame, but will allow a great deal of smoke penetration.

STANDARD FIRE DOORS:
1. Overhead rolling.
2. Horizontal and vertical sliding.
3. Single and double swinging.

EXIT DOORS in a theater should swing out in the direction of the street so as the exits are more readily seen.

AUTOMATIC FIRE DOORS: normally remain open but will close when heat actuates their closing device.

SWINGING FIRE DOOR: used on stair enclosures.

FIRE WALLS: erected to prevent the spread of fire.

FIRE PARTITION is a partition which serves to restrict the spread of fire but does not qualify as a fire wall. (rated for 2 to 4 hours)

BEARING WALL: capable of supporting a vertical load such as a floor, roof, in addition to its weight. **N.F.P.A.** table #5 is used for the determination of wall and opening protection.

EXIT ACCESS: the means of egress which leads to an exit.

EXIT: the portion of escape, doors, walls, etc., which provide a protected path to exit discharge.

EXIT DISCHARGE: the portion of travel from the exit to a public way.

MEANS OF EGRESS: continuous, unobstructed way of exit travel from any point in a building or structure to a public way.

The three separate parts of **"MEANS OF EGRESS"**:
    1. Exit access.
    2. Exit.
    3. Exit discharge.

THREE CLASSES OF OPENINGS:
    1. Class A : Separating buildings or dividing buildings into fire areas.
    2. Class B : Vertical openings.
    3. Class C : Corridor - room partitions.

CAPACITY OF EXITS is used to establish a consistency of elevation time on the basis of the rate of travel through a door of 60 persons per minute and down a stairway of 45 persons per minute

PANIC HARDWARE on doors, the pressure is not to exceed 15 LBS.

PEOPLE MOVEMENT is the movement of occupants and firefighters.

CLASSIFICATION OF OCCUPANCIES:
1. A = Assembly.
2. B = Business.
3. E = Educational.
4. H = Hazardous.
5. I = Institutions.
6. M = Carports - Fences.
7. R = Residences.

The **TYPE** of occupancy of a building determines the degree and nature of the hazard, along with the fire potential that may exist.

OCCUPANCY HAZARDS are equally as important as construction features in relation to the study of ventilation.

LIFE SAFETY CODE requires that in assembly occupancies that the main exit be sized so as to handle at least one half the occupant load. Main exit also serves as entrance.

The **BEST** inspection approach is good public relations, education, and then enforcement.

NEVER inspect a building without the permission of the occupant.

INSPECTION: the close and critical examination by a competent authority.

Fire prevention inspection should be conducted at **IRREGULAR** intervals so that the inspector may see the inspected establishments in their normal conditions.

NATIONAL FIRE PREVENTION WEEK is declared by **PRESIDENTIAL PROCLAMATION.**

A fire prevention inspection should be the most thorough and complete by the **FIRE DEPARTMENT** than by any other agency.

FIRE HAZARD: any material, condition, or act that will contribute to the start of fire or will increase the severity or extent of the fire.

FIRE CAUSE involves three controllable conditions:
1. Fuel supply.
2. Heat source.
3. The hazardous act.

SPECIAL FIRE HAZARD: fire hazard arising from the processes or operations that are peculiar to the individual occupancy.

COMMON FIRE HAZARD: The condition that is likely to be found in most occupancies and is not generally associated with any specific occupancy, process, or activity.

TARGET HAZARD: A condition, facility, or process which could produce or stimulate a fire that would involve a possible large life loss, a possible large fire loss, a large concentration of materials of high monetary value.

Approximately **99%** of the **SUCCESS** of a fire prevention program depends on **VOLUNTARY** actions of building occupants.

INSPECT DURING FIRE INSPECTIONS:
1. Waste disposal.
2. Trash receptacles.
3. Trash collection points.

RESPONSIBILITY rest with the **FIRE CHIEF** for determining of fire cause and origin.

ARSON is a crime which is hard to secure evidence because the evidence is usually consumed.

MOST IMPORTANT FACTOR in prevention of school fires, is good housekeeping.

SCHOOL INSPECTIONS should be conducted monthly.

Test **HOME** smoke detectors monthly.

SMOKE DETECTORS have the potential to reduce home fires by **40%**. The best spot for **SMOKE DETECTORS** in a home is in the hallway, outside the bedrooms.

E.D.I.T.H. = Exit drills in the home.

Two thirds of all **FIRE DEATHS** occur in the home.

LEADING CAUSE of fire is smoking.

FLAMMABLE LIQUIDS = liquids having flash point below **100 DEGREES F**.

COMBUSTIBLE LIQUIDS = liquids having flash point of
100 DEGREES F OR HIGHER.

CLASSES OF FLAMMABLE LIQUIDS:
1. Class I   : FP = less than 73 degrees F.
2. Class IA : FP = less than 73 degrees F.
             BP = less than 100 degrees F.
3. Class IB : FP = less than 73 degrees F.
             BP = greater than 100 Degrees F.
4. Class IC : FP = between 73 and 99 degrees F
   (1-4 FP = flash point, BP = boiling point)

CLASSES OF COMBUSTIBLE LIQUIDS:
1. Class II   : FP = 100 to 140 degrees F.
2. Class III : FP = greater than 140 degrees
3. Class IIIA : FP = 140 to 199 degrees F.
4. Class IIIB : FP = 200+ degrees F.

HAZARDOUS MATERIALS

SMOKES DEADLY TRIO:
1. Carbon Monoxide.
2. Hydrogen Sulfide.
3. Hydrogen Cyanide.

Most fire related **DEATHS** occur from **CARBON MONOXIDE**.

CARBON MONOXIDE is produced by each fire but is produced in larger quantities in fires of a **SMOLDERING** nature.

The darker the smoke the higher the content of **CARBON MONOXIDE**.

CARBON MONOXIDE = incomplete combustion; firefighters fighting cellar fires should be especially alert to the possible presence of **CARBON MONOXIDE** for this reason.

An indication of **INCOMPLETE COMBUSTION** is smoke and toxic gases.

CARBON MONOXIDE is lighter than room air.

The blood system absorbs **CARBON MONOXIDE** 210 times more readily than oxygen.

CARBON MONOXIDE is the chief danger in most fire gases. It combines with hemoglobin **210 TIMES** more readily with oxygen. It robs the blood of oxygen. **1.3%** will cause unconsciousness in 2 or 3 breaths and death in a few minutes.

Concentrations of more than .05% of **CARBON MONOXIDE** are dangerous.

BACKDRAFT is a rapid, almost instantaneous combustion of flammable gases, Carbon particles, and Tar balloons emitted by burning materials under conditions of insufficient Oxygen. Normally in confined spaces.

AIR IS COMPOSED OF:
1. Oxygen.
2. Nitrogen.
3. Traces of, Hydrogen, sulphur, and Phosphorus.

NORMAL AIR = 21% Oxygen and 79% Nitrogen.

HYDROGEN SULFIDE is prevalent in industrial fires of wool or rubber.

HYDROGEN SULFIDE is a skin irritant.

HYDROGEN SULFIDE is colorless.

HYDROGEN SULFIDE has the odor of "rotten eggs".

HYDROGEN SULFIDE is heavier than air.

HYDROGEN SULFIDE is produced by the decomposition of organic materials and the actions of acids on metallic sulfides.

HYDROGEN SULFIDE attacks the nervous system.

HYDROGEN SULFIDE is produced by the incomplete combustion of organic materials containing:
1. Wool.
2. Animal hides.
3. Animal Meat.
4. Animal hair.

HYDROGEN CYANIDE has a bitter almond taste.

HYDROGEN CYANIDE has a **SYNERGISTIC** effect when it is combined with carbon monoxide.(combined action)

HYDROGEN CYANIDE is colorless.

HYDROGEN CYANIDE has a tendency to increase the heartbeat, thus making you breathe more carbon monoxide gas in physical and stressful situations.

HYDROGEN CYANIDE is not likely to be produced in dangerous quantities in most fires. Will be present in buildings that are being fumigated.

CARBON DIOXIDE overstimulates the rate of breathing. A **10%** ratio may cause death in a few minutes, also breathing will increase which increases the intake of toxic gases.

CARBON DIOXIDE is **1.5** times as heavy as air.

CARBON DIOXIDE the maximum amount that people can withstand without losing consciousness is **9%**.

CARBON DIOXIDE will supply oxygen to magnesium fires.

CARBON DIOXIDE"S hazard is that it stimulates breathing while in contaminated atmospheres.

CARBON DIOXIDE and **CARBON MONOXIDE** are both odorless and colorless.

NITROGEN DIOXIDE is formed with oxides of nitrogen during decomposition and combustion of **CELLOSE NITRATE** and fires with **AMMONIUM NITRATE** and other **INORGANIC NITRATES**.

NITROGEN DIOXIDE has a reddish brown color.

ACROLEIN is produced during combustion of petroleum products.

ACROLEIN: 10 parts per millon is lethal in a short time.

AMMONIA is formed during the burning of material containing Nitrogen.

AMMONIA is two gaseous elements: Nitrogen and Hydrogen.

CHLORINE is 2.5 times as heavy as air.

CHLORINE causes spontaneous combustion in turpentine and ammonia.

TURPENTINE = a lower hazard than lacquer, shellac, and varnish.

To locate a **CHLORINE** leak: wrap a rag on the end of a stick and saturate in Ammonia water, when this is held near a **CHLORINE** leak a white smoke cloud will appear.

HYDROGEN = .1 vapor density. (Air = 1)

HYDROGEN is the lightest gas known.

FREON is non-flammable and slightly toxic.

LAMP BLACK is formed by burning low grade heavy oils with insufficient air.

METHYL-BROMIDE is non-flammable, but toxic.

CARBON BLACK is formed ny the decomposition of Acetylene or by the incomplete combustion of Natural gas.

CARBON TETRACHLORIDE vaporizes to form a heavy non-flammable gas.

The gases released from **STORAGE BATTERIES** are Hydrogen and Oxygen.

CARBON DISULFIDE is given off from the burning rubber insulation on electrical wiring.

CYANIDE is highly soluble in water.

CYANIDE GAS is very poisonous.

L-P GASES have high density, are heavier than air.

L-P GASES = approximately twice the weight of air:
 1. Propane = 1.5 the weight of air.
 2. Butane = 2.05 the weight of air.

GASES expand indefinitely.

FLAMMABLE GASES do not have flash points.

ASPHYXIA is the lack of Oxygen.

CHLORATES give off high amounts of Oxygen.

OXYGEN will not burn, but is a strong **SUPPORTER** of combustion.

HALOGENS are salt producing chemicals.

SOME EXAMPLES OF HALOGENS:
 1. Fluorine.
 2. Chlorine.
 3. Bromine.
 4. Iodine.

UNSTABLE MATERIALS are those which will polymerize, decompose, condense, or become self-reactive when exposed to air, water, heat, shock, or pressure.

SPONTANEOUS = slow oxidation.

OXIDATION a chemical reaction in which Oxygen combines with other substances.

OXIDATION any chemical reaction in which Electrons are transferred.

OXIDATION and reduction always occur simultaneously, and the substance which gains the Electrons is called the **OXIDIZING AGENT**.

OXIDATION always produces heat.

REDUCTION is a chemical process in which the Oxygen is removed from a compound or environment.

MAGNESIUM and **SODIUM** are reducing agents.

GASOLINE is considered a stable material.

SPONTANEOUS HEATING is heating due to chemical or bacterial action in a combustible material.

SPONTANEOUS IGNITION is ignition due to chemical reaction or bacterial action in which there is a slow oxidation of organic compounds until the material ignites; usually there is sufficient air for oxidation but not enough ventilation to carry heat away as it is generated.

COAL is subject to spontaneous heating and ignition, except for high grade Anthracite.

YELLOW PHOSPHORUS reacts violently to air.

SUBLIMATION: evaporation or release of vapors from a solid without going through the liquid phase.

CAMPHOR and **NAPHTHALENE** change from a solid to a vapor without passing through the liquid phase. This is called **SUBLIME**.

Chemicals are **MORE DANGEROUS** during handling than when stored, because they are exposed to changing conditions.

The first principle of chemical storage is **SEGREGATION**.

CHEMICAL COMPOSITION of most ordinary combustible solids consist primarily of:
    1. Carbon.
    2. Hydrogen.
    3. Oxygen.

NITRITES:
1. Inorganic Peroxides.
   a) Sodium.
   b) Potassium.
   c) Strontium.

NITRATES:
1. Sodium.
2. Potassium.
3. Ammonium.

NITRATES will cause fire to intensify and they also release toxic oxides.

DIFFERENT TYPE OF CHEMICAL REACTIONS:
1. Water and air reactive chemicals.
2. Unstable chemicals.
3. Combustible chemicals.
4. Corrosive chemicals.

WATER AND AIR REACTIVE CHEMICALS:
1. Carbides.
2. Alkalies (caustic)
   a) Anhydrides b) Aluminum Triackls.
3. Phosphorus, Red Phosphorus.
4. Hydrides, Hydrosulfite (Sodium)
5. Oxides.
6. Coal and Charcoal.

WATER AND AIR REACTIVE CHEMICALS: EXAMPLES
1. Lye
2. Acetic Anhydride.
3. Sodium Carbide.
4. Charcoal, Coal.
5. Lithium Hydride.
6. Quicklime.

UNSTABLE CHEMICALS: EXAMPLES
1. Ethylene.
2. Nitro Methane.
3. Oxides, Organic Peroxides.
4. Acetaldehyde.
5. Hydrogen Cyanide.

COMBUSTIBLE CHEMICALS: EXAMPLES
1. Lamp Black.
2. Organic Peroxides.
3. Carbon Black.
4. Naphthalene.
5. Sulfides, Sulphur.

CORROSIVE CHEMICALS: EXAMPLES
1. Sulfuric Acid.
2. Perchloric Acid.
3. Inorganic Acid.
4. Nitric Acid.
5. Hydrochloric and Hydrofluoric Acid.

COMBUSTIBLE METALS: EXAMPLES
1. Potassium.
2. Calcium.
3. Lithium.
4. Hafnium.
5. Magnesium.
6. Titanium.
7. Sodium.
8. Zinc and Zirconium.

Most flammable liquids have a **SPECIFIC GRAVITY** of less than 1, water = 1 therefore they will float on water.

CHEMICAL REACTIONS double their rate with each **18% RISE** in temperature.

FLAMMABLE EXPLOSIVE LIMITS have upper and lower limits expressed in percentage of vapor within air.

Most intense explosion will occur with vapor at **MID-RANGE** of products explosive range.

CARBON MONOXIDE and air mixture has an **EXPLOSIVE RANGE** of 12.5% to 74% by volume.

ACETYLENE has the widest **EXPLOSIVE RANGE** of any gas.

To **STABILIZE** acetylene for compression in tanks, **ACETONE** is added.

BRISANCE is the effect of shattering, explosive (measurement).

BLEVE: Boiling Liquid Expanding Vapor Explosion.

Signs of **BLEVE:**
1. Direct heat on vessel.
2. System is venting.

The **WIDER** the flammable range of a gas then the more dangerous the gas.

PRESSURE of gas over liquid in a vessel is dependent upon the amount of the product in the vessel and the temperature of the liquid and gas.

ENDOTHERMIC REACTION: a process or change that absorbs heat and requires it for initiation and maintenance.

DUST CLOUDS of most metals are explosive.

CONCENTRATION of dust is expressed as a percentage.

As dust particle size **DECREASES** the explosion potential increases.

Impurities **REDUCE** the potential of dust explosion.

Lowering the Oxygen content **REDUCES** the potential of dust explosion.

The more finely divided the particles of **COMBUSTIBLE METALS** are, the easier they are to ignite.

Their are two classifications of **COMBUSTIBLE DUST:**
    1. Class A = Metals.
    2. Class B = Non-metals.

LETTER DESIGNATIONS OF SOME HARMFUL CHEMICALS:
    1. CO = Carbon Monoxide.
    2. CO2 = Carbon Dioxide.
    3. HCI = Hydrogen Chloride.
    4. HCN = Hydrogen Cyanide.

ATOMIC ENERGY COMMISSION (AEC) is the Federal agency whose primary function is to regulate and control the uses of radioactive materials.

CHEMTRIC = MCA : Manufacturing Chemist Association.

CHEMTRIC toll-free telephone number = 800-424-9300.

D.O.T. = Department of transportation.

D.O.T. label for **EXPLOSIVES:**
    1. Square shape.
    2. Red color.
    3. Black lettering.

D.O.T. PLACARDS the number zero would equal the least hazardous, totally stable.

D.O.T. label for **FLAMMABLE SOLIDS:**
    1. Diamond shape.
    2. Yellow color.
    3. Black lettering.

D.O.T. label for **NON-FLAMMABLE COMPRESSED GAS:**
    1. Diamond shape.
    2. Green color.
    3. Black lettering.

D.O.T. label for **POISONOUS ARTICLES:**
    1. Red and white colors.
    2. Red lettering.

D.O.T. PLACARD the number of four would equal the most hazardous, the least stable.

PH SCALE:
    1. Above 7.0 = bases : usually solid.(alkalis)
    2. Below 7.0 = acid : usually liquid.
    3. At   7.0 = Neutral.

FIREFIGHTING/FIRE BEHAVIOR

The **PRIMARY FUNCTION** of an engine company is to **OBTAIN** and **DELIVER WATER**.

SEQUENCE OF FIRE FIGHTING:
1. Locate the fire.
2. Confine the fire.
3. Extinguish the fire.

The strategy of **THREE - PRONGED** or **THREE POSITION ATTACK** is necessary to control and extinguish any fire of any magnitude.

The three positions of **THREE - PRONGED** or **THREE POSITION ATTACK** are:
1. Exposures.
2. Avenues of fire spread.
3. The seat of the fire.

About **95%** of all fires are extinguished with one line or less.

PRIMARY duty of an officer of a pumper company, after assignment to tactical position is, **"SETTING UP THE PUMPER"** at a suitable hydrant or water source.

Protection of life is the **NUMBER ONE PRIORITY** of the fire department.

AMOUNT OF WATER for a fire depends on the amount of heat generated.

For covering all points of a fire remember : **"FRONT** and **REAR, OVER** and **UNDER,** and **COVER ALL EXPOSURES"**.

POOR hose lays = low capacity.

1 1/2" hose (50 GPM to 125 GPM) is the **BEST** size of hose for **INITIAL STRUCTURAL FIRE,** but not for buildings with large areas.

For fire in an average small building, the first two pumpers should be able to apply **400 GPM** as a minimum promptly on arrival.

LARGEST NUMBER of fire streams that first in engine company can stretch and operate is **TWO**.

MINIMUM water requirement for sufficient fire streams for dwelling fire is **500 GPM** = two 1 1/2" lines at 75 GPM to 125 GPM plus one 2 1/2" line at 200 GPM to 250 GPM.

25,000 CUBIC FEET is the maximum fire area for a manually applied hose stream, covered by the first alarm response.

Small fire streams = **MANEUVERABILITY**.

85% of all fires are extinguished with **BOOSTER** or 1 1/2" **LINES**.

HEAVY STREAM placement should include coverage of exposures as well as cooling of the fire, if possible.

If there is an alleyway or driveway between fire and exposure, an important **DEFENSE POSITION** against lateral spread of fire is to use heavy streams backed up by handlines.

Officer in charge at large lumber yard fire should place **HEAVY STREAMS** from deck pipes and towers on the leeward side of the fire and hand lines on the sides and the rear.

When making an **INDIRECT ATTACK** on a fire in the interior of a building, the best attack is from positions outside the building.

When determining size of lines to be laid at a fire, the most significant factor is the **EXTENT** of the fire and the **COMBUSTIBILITY** of the materials involved.

SECONDARY COMMAND POST (staging) in high rise fire should be second floor below the fire.

FIRE FLOOR: the floor or story of a building on which a fire is burning.

Use standpipe system instead of stretching hoselines up stairways above the **THIRD FLOOR**.

MAXIMUM story to advance hoselines from ground level is to the **THIRD FLOOR**.

In a multistory building with the fire on the 4th floor, the best procedure for the engine company is to **CONNECT ITS LINES TO THE STANDPIPE SYSTEM.**

Connect hose line to standpipe system on the **FLOOR BELOW THE FIRE.**

In high rise buildings **FIRE LOSSES** vary **GEOMETRICALLY** accordingly to the time fire burns.

At large fires there is a need for a **WATER SUPPLY OFFICER.**

A **SECOND ALARM** response should equal the first alarm response.

INITIAL RESPONSE to all structures of moderate fire hazard should be capable of immediately applying **400 GPM**, backed up by adequate pumping capacity and water supply. For larger and higher hazards 1000 GPM is needed.

FIRE FLOW IN A CITY IS DEPENDENT UPON:
1. The size of the most congested area.
2. The hazards of the most congested area.
3. The structural conditions of the most congested area.

BLITZ ATTACK equals speed.

THEORY can be best aid for lack of experience.

PRE-FIRE PLANS are the best substitute for lack of experience.

SIZE-UP: The mental evaluation made by the fire officer in charge, which enables him to determine a course of action; it includes such factors as time, location, nature of occurrence, life hazard, exposure, property involved, nature and extent of the fire, available water supply and other fire fighting facilities. **SIZE-UP** is a report usually via radio, giving existing conditions of an emergency.

Four stages of **SIZE-UP:**
1. Anticipating the situation.
2. Gathering the facts.
3. Evaluating the facts.
4. Determining the procedures.

MOST IMPORTANT consideration of **SIZE-UP** is location.

SIZE-UP is more important at large fires.

SIZE-UP is continuous throughout the entire fire fighting operation.

FIRE TACTICS: various maneuvers that can be employed in a strategy to successfully fight a fire.

FIRE STRATEGY: the plan of attack on a fire.

FIRE STRATEGY: should make prime use of equipment and personnel, and take into consideration fire behavior, he nature of the occupancy, environmental conditions, and other factors.

TASK FORCE concept = command.

SPEED OF DISCOVERY = most important to fire loss.

MANPOWER is most critical at early stages of fire.

FIRE SCENE: the fire ground.

FIRE SCIENCE the knowledge concerning the behavior, effects, and control of fire.

FIRE SPREAD: the involvement and migration of fire across surfaces. (flame spread)

FLASHOVER: a fire continues to burn all the contents gradually heating to their ignition temperatures.

FIRE STORM: a violent, convective atmospheric disturbance caused by large intense fires which tend to suck all the available air into the fire.

FIRE TRIANGLE: a three sided figure representing the three of the four factors necessary for combustion.
1. Oxygen.
2. Heat.
3. Fuel.

FIRE TETRAHEDRON: the four elements required by a fire:
1. Fuel.
2. Heat.
3. Oxygen.
4. Uninhibited chain reaction.

FIRE WIND: wind caused by an intense fire that consumes oxygen from the atmosphere which creates a partial vacuum and causes the movement of more air towards the fire.

FIRE BURNS IN TWO MODES:
1. Flaming mode   = Tetrahedron the second phase is free burning.
2. Smoldering mode = Fire triangle the third phase is smoldering.

STAGES OF FIRE:
1st = oxygen at 21%, fire at 1000 degrees F, room at 100 degrees F.
2nd = oxygen at 21% to 15%, fire and room at 1300 degrees F.
3rd = oxygen below 15%, fire and room at 1000 degrees F.

FIRST STAGE OF FIRE = smoldering, incipient phase.

SECOND STAGE OF FIRE = flame producing phase.

THIRD STAGE OF FIRE = smoldering phase.

SMOLDERING PHASE of fire = decrease in heat generation.

BACKDRAFT CHARACTERISTICS:
1. Smoke under pressure.
2. Dense grayish, yellowish smoke.
3. Puffing smoke from cracks, moving up rapidly.
4. Confinement of excessive heat.
5. Sweating windows, hot to the touch, and dark in color.
6. Muffled sounds.
7. No visible flame.
8. Rapid movement of air inward when opening is made.

PRODUCTS OF COMBUSTION:
1. Fire gases; Oxygen, Hydrogen, and Carbon.
2. Flame.
3. Heat.
4. Smoke.

COMBUSTION: a rapid exothermic oxidation process accompanied by continuous evolution of heat and usually light.

EXOTHERMIC HEAT: gives up heat.

ENDOTHERMIC HEAT: absorbs heat; chemical reaction.

HEAT: proportional energy created by the motion of molecules that can be transferred from one body to another by radiation, conduction, and convection.

TRANSMISSION OF HEAT:
1. Conduction; direct heat contact.
2. Radiation; in all directions where matter does not exist, such as air.
3. Convection; by air currents usually in an upward direction.
4. Direct flame contact.

DAYS LEAST FAVORABLE TO FIRES:
1. Relative humidity less than 40%
2. Precipitation less than .01 the day of fire
3. Precipitation less than .01 three days prior to fire.
4. Maximum wind speed is no more than 13 MPH.

WEATHER: the general state of the atmosphere at a specific time and place, with respect to temperature, cloudiness, moisture, etc.

FIRE DEVIL: a small, burning cyclone that occurs usually during forest and brush fires but can also occur in free burning structural fires. Fire devil will form a vortex as heated gases from a fire rise and cooler air rushes into the resulting areas of low pressure.

FIRE DANGER RATING AREA: is a geographical area that is comparably consistent with respect to climate, fuels, and topography, for which the fire danger is constant.

FIREBREAK: a natural or constructed barrier that stops the spread of a fire or provides a control line from which to work.

CONFLAGRATION HAZARD: is a close group of structures that are subject to rapid fire spread.

Most common spread of fire in buildings is **UNPROTECTED VERTICAL OPENINGS.**

Fires are most frequently spread by **CONVECTION.**

Firefighting within a structure from several different directions can be very difficult mainly because the **SMOKE** and **HEAT** will be driven from one direction to another.

The most significant factor for spreading fire in structures are **STAIRS** and **SHAFTS**.

HORIZONTAL EXTENSION OF FIRE MAY OCCUR:
1. Through wall openings by direct flame contact.
2. Through open space by radiated heat or by convected air currents.
3. Through walls and interior partitions by direct flame contact.

DOWNWARD EXTENSION OF FIRE MAY OCCUR:
1. Through floors by direct flame contact.
2. Through ceilings by direct flame contact.

VENTILATION: a planned and systematic release and removal of heated air, smoke, and gases from a structure and the replacement of these products of combustion with a supply of cooler air.

A awareness of **VENTILATION** is of preeminent importance, second only to the application of an extinguishing agent.

FIRE VENTILATION will provide a better condition for breathing and heat, but will not remove all the hazards or the dangerous gases.

SAFETY OF THE OCCUPANTS is the first consideration of ventilation.

ARTIFICIAL VENTILATION at both the floor and ceiling will provide greater safety, when removing extensive amounts of flammable vapors.

VENTILATION AS DIRECTLY OVER THE FIRE as possible is the best rule of thumb for selecting the point to open the roof.

Firefighters should be aware of **INTERNAL EXPOSURES** and **EXTERNAL EXPOSURES** when performing **HORIZONTAL VENTILATION**.

VENTILATE AT THE BASE OF THE BUILDING for a fire involving a refrigeration plant.

When ventilating a building with a hole on the roof, doors in the lower portion of the building should be left **OPEN** so as to let in cool air.

During a slow smoldering fire within a small area of a structure, dense smoke should be cleared by **VENTILATION** so that the fire may be located.

BASEMENT VENTILATION usually requires **MECHANICAL** ventilation.

During ventilation on the roof of a building involved in fire, a **SPONGY ROOF** may indicate that the structural members have been weakened.

The **SPECIFIC HEAT** of a material describes the materials ability to absorb heat.

Common combustible materials will not burn if the Oxygen percentage drops below or at **15%**.
When a Firefighter enters a building involved in fire, he can expect to find the **PRODUCTS OF COMBUSTION**, depending on what is burning.

PROTECTIVE BREATHING EQUIPMENT should always be donned when a Firefighter enters a contaminated atmosphere, as it is extremely difficult to know what gases are present.

Smoke normally **RISES** up from fire because cooler, heavier air displaces the lighter warmer air.

MUSHROOMING: the condition when the heat and gases spread out laterally at the top of a structure.

BOYLE'S LAW: if the temperature remains constant, the volume of gas varies inversely with the pressure.

In a fire building supported by exposed steel columns and girders, the steel can **CONDUCT** or **RADIATE** heat to combustible materials not in direct contact with the fire.

The most difficult of all fires of their size to combat are **BASEMENT FIRES.**

Burning **LIQUID FUEL MATERIALS** generally give off dense black smoke.

EXTINGUISHING SYSTEMS

Portable fire extinguisher are classified according to their **INTENDED USE**.

Information for portable fire extinguisher can be found in **N.F.P.A. STANDARD #10**.

FIRE EXTINGUISHER RATINGS AND COLOR CODES:
    Class A fires = Ordinary combustibles:
        Green triangle.
    Class B fires = Flammable liquids:
        Red square.
    Class C fires = Electrical:
        Blue circle.
    Class D fires = Metal:
        Yellow star.

AGENTS FOR FIRE EXTINGUISHING SYSTEMS:
1. Dry chemical.
2. Carbon dioxide.
3. Foam.
4. Halons.
5. Water spray.

WATER as a **COOLING AGENT** for flammable liquids:
1. Cuts off release of vapor from the surface of a high flash point oil, thus extinguishes the fire.
2. Protects firefighters from flame and radiant heat.
3. Protects flame exposed surfaces.

WATER as a **MECHANICAL TOOL** for flammable liquids:
1. Controls leaks.
2. Directs the flow of the product to prevent it's ignition, or to move the fire to an area where it will do less damage.

WATER as a **DISPATCHING MEDIUM** for flammable liquids:
1. Will float oil above a leak in a tank or during a fire.
2. Will cut off the fuels escape route by pumping it into a leaking pipe before the leak.

WATER is the most important extinguishing agent because of its physical characteristics, its universal availability, and because of it's low cost.

WATER will absorb heat to a much greater extent than any other material that is easily available.

WATER has a lack of opacity, thus it has little ability to prevent the passage of radiant heat.

Adding **WETTING AGENTS** to water increases the waters heat absorption ability.

WETTING AGENTS reduce the surface tension of water.

WETTING AGENTS increase the waters penetration ability.

A gallon of water produces a maximum of **200 CUBIC FEET OF STEAM.**

The formation of steam by water that has been applied to the burning source will temporarily cause an inert gaseous water zone (steam) in and around the burning zone.

WET-WATER solutions will foam easily, and the temporary foam will control and extinguish class B fires better than ordinary water.

BASIC TYPES OF FOAM FOR FIREFIGHTING:
    1. Chemical foam.
    2. Mechanical foam.

SUBSURFACE FOAMS injected or intermixed with the liquids that are involved are called flouroprotein.

WETTING AGENTS reduce the surface tension of water which increases the penetrability.

SURFACE TENSION is the resistance to penetration possessed by the surface of a liquid.

DIRECT foam against the far side of tank shell when extinguishing an oil tank fire.

The chief purpose of a **STABILIZER** in foam extinguisher is to prevent the rapid breakdown of the carbon dioxide (CO_2) bubbles.

AFFF FOAM RATIO:
1. Hydrocarbons (petroleum products) = 3% foam to 97% water.
2. Polarsolvents (water soluble) = 6% foam to 94% water.

FOAM is the most effective extinguishing agent for oil fires, because it is lighter than oil and remains on top.

WATER FOG is not capable of extinguishing flammable liquid fires with a flash point below 100 degrees F

SPRAY FIRE STREAMS IN CONTRAST TO SOLID STREAMS:
ADVANTAGES:
1. Absorbs more heat, more rapidly.
2. Covers a greater area with water.
3. Uses less water.
DISADVANTAGES:
1. Requires higher discharge pressure.
2. Has a shorter reach.
3. Has less penetration.
4. Has less cooling effect in subsurface areas such as charred wood, etc.

VISCOUS WATER (thickened water:
Advantages
1. Sticks to burning fuel.
2. Spreads itself in continuous coating.
3. Thicker than plain water.
4. Absorbs more heat.
5. Projects longer and higher straight streams
6. Seals fuel from oxygen after drying.
7. Resist wind drift. (as from aircraft in forest fires).

LIGHT WATER (fluorinated surfactant), is useful in obtaining quick knockdown of flammable liquid fires, and in providing a vapor sealing effect for reducing subsequent flashover of fuel vapors exposed to lingering.

CONVENTIONAL FOAM is formed by the reaction of alkaline salt solution in acid salt solution in the presence of a foam stabilizing agent, and mechanical or air foam formed by turbulent mixing of air with water containing foam forming agents.

HIGH EXPANSION FOAM is for **CLASS "A" AND "B"** fires and suited as a flooding agent in confined spaces.

LOW EXPANSION FOAMS can be used to good effect when the temperature of the bulk of a contained liquid does not exceed **250 DEGREES F**.

ORDINARY FOAM IS BROKEN DOWN BY:
1. Common alcohols.
2. Aldehydes.
3. Ethers.

ALCOHOL FOAM is recommended for all water soluble flammable liquids except for those that are only very slightly soluble.

The **MOST IMPORTANT** use of **FOAM** is in fighting fires in petroleum hydrocarbons (gasoline) with high vapors and low flash points.

FOAM should never be used on energized electrical equipment.

PROPORTIONER is the device that is used to inject the correct amount of foam concentrate into the water stream so as to make the foam solution the correct proportion.

FIVE GALLONS of foam powder = **400 GALLONS** of foam.

2A WATER fire extinguisher will cover a **100 FOOT** area.

The principle extinguishing agent of **SODA** and **ACID EXTINGUISHER** is the **WATER** content.

The most widely used water soluble freezing point depressant in fire equipment is **CALCIUM CHLORIDE**.

SODA ACID EXTINGUISHER have Bicarbonate Soda and Sulfuric Acid as ingredients.

REGULAR DRY CHEMICAL EXTINGUISHER are Bicarbonate base powder for Class B and Class C fires.
(Sodium Bicarbonate and Potassium Bicarbonate)

MULTIPURPOSE DRY CHEMICAL EXTINGUISHER are Ammonium Phosphate base powders for Class A, B, and C fires.
(Monoammonium Phosphate, also Barium Sulfate)

DRY CHEMICAL EXTINGUISHER are a mixture of Sodium Bicarbonate, Potassium Bicarbonate, or Ammonium Phosphate.

DRY CHEMICAL EXTINGUISHER EXTINGUISH BY:
1. Smothering.
2. Cooling.
3. Radiation shielding.
4. Chain breaking; a reaction in the flame may be the principle cause of extinguishment.

EXTINGUISHMENT BY SEPARATION cannot be accomplished with burning materials that contain their own oxygen supply, such as Cellose Nitrate.

EXTINGUISHMENT BY SEPARATING the oxidizing agent from the fuel is accomplished by blanketing or smothering a fire.

CHEMICAL COMPOSITION of most ordinary combustible solids consist of primarily: Carbon, Hydrogen, and Oxygen.

Use **DRY CHEMICAL EXTINGUISHER** on LP gas fires.

DRY CHEMICAL EXTINGUISHER should not be used where relays or other delicate electrical contacts are located such as telephone exchanges.

DRY CHEMICAL EXTINGUISHER will not extinguish fires in materials that supply their own Oxygen.

MULTIPURPOSE DRY CHEMICAL EXTINGUISHER, extinguish by the Ammonium Phosphate decomposing and leaving a sticky residue on the burning material, this seals the material off from Oxygen.

5 LBS of **DRY CHEMICAL** is as effective as **10 LBS** of **CARBON DIOXIDE.** (CO_2)

MULTIPURPOSE DRY CHEMICAL EXTINGUISHER may be used on Class A fires.

FOAM EXTINGUISHER must be protected from freezing.

Extinguish wetted Carbide with **DRY POWDER**.

CARBON DIOXIDE FIRE EXTINGUISHER extinguish by cooling with the rapid expansion of liquid to gas producing a refrigerating effect.

15 LBS CARBON DIOXIDE EXTINGUISHER range = 8'-10'.
CARBON DIOXIDE EXTINGUISHER are not effective on fires involving reactive metals.

CARBON DIOXIDE EXTINGUISHER are under a pressure of between **800 PSI** and **900 PSI**.

CARBON DIOXIDE EXTINGUISHER with CO2 under pressure, the purpose of the horn dispenser is to avoid entraining air as the contents pass through the small orifice at high velocity.

CARBON DIOXIDE EXTINGUISHER suppress chain reaction of combustion.

1 LB of **CARBON DIOXIDE** liquid will produce **8 CUBIC FEET OF GAS** at atmospheric pressure.

CARBON DIOXIDE EXTINGUISHER are not effective on fires involving chemicals containing their own Oxygen supply, such as Cellulose Nitrate.

CARBON DIOXIDE EXTINGUISHER are recommended for stove fires.

CARBON DIOXIDE EXTINGUISHER are recommended for fires involving energized electricity.

Use care with **CARBON DIOXIDE EXTINGUISHER**, mainly because of the possibility of **RE-FLASH**.

CARBON TETRACHLORIDE vaporizes to form a heavy non-flammable gas.

PORTABLE FIRE EXTINGUISHER WHICH ARE CONSIDERED OBSOLETE:
    1. Soda Acid.
    2. Carbon Tetrachloride.
    3. Loaded Stream.
    4. Cartridge-operated Water.
    5. Inverting Foam.

COMBUSTIBLE METAL FIRES (Class D) are extinguished by a number of approved extinguishing agents; Powders, and Dry Powders. Each for specific metals.

HALOGENATE EXTINGUISHING agents work by:
    1. Vaporizing liquid fire extinguishing agents.
    2. Chain Breaking.
    3. By providing non-flammability and extinguishing characteristics.

HALON EXTINGUISHING AGENTS:
1. Fluorine.
2. Chlorine.
3. Bromine.
4. Iodine.

SPRINKLER RATINGS:
Ordinary   - No color = 135 - 170 degrees F.
Intermediate   - White   = 175 - 225 degrees F.
High   - Blue   = 250 - 300 degrees F.
Extra high   - Red   = 325 - 375 degrees F.
Very extra high - Green   = 400 - 475 degrees F.
Ultra high   - Orange   = 500 - 575 degrees F.

PIV = POST INDICATOR VALVE.
SPRINKLER COVERAGE according to occupancy hazard, with 1/2 inch orifice:
Light   = 130 square feet to 168 square feet.
Ordinary   = 130 square feet.
Extra high = 90 square feet.

WATER SUPPLIES FOR SPRINKLER SYSTEMS:
1. Public water works systems.
2. Public and private supplies.
3. Gravity tanks. (minimum of 5000 gallons)
4. Pressure tanks. (minimum of 4500 gallons)
5. Fire pumpers.
6. Fire department connections.

AUTOMATIC SPRINKLERS should be installed in warehouses of Type I or II, but not of Type III.

The **MAIN** reason that a sprinkler system should be installed inside of buildings is to extinguish fires in their early stages.

AUTOMATIC SPRINKLERS under normal situations are highly efficient and dependable. If an explosion takes place, sometimes this will cause the sprinkler head and piping to be so badly damaged that they will be ineffective.

AUTOMATIC SPRINKLERS are the most effective safeguard against loss of life by fire. Psychological as well as physical, by minimizing panic.

TOTAL SPRINKLER SYSTEMS (piping and devises) are hydrostatically tested at not less than **200 PSI FOR 2 HOURS** or 50 PSI in excess of the maximum static pressure when it is above 150 PSI.

DRY PIPE SPRINKLER SYSTEMS, the piping contains air under pressure, 15 PSI to 20 PSI and no water in the piping.

PRE-ACTION SPRINKLER SYSTEMS are activated at the water supply valve, not at the sprinkler head, like in the standard type dry sprinkler system.

WET PIPE SPRINKLER SYSTEMS are fully charged with water.

DELUGE SPRINKLER SYSTEMS wet down the entire area by admitting water to sprinklers that are open at all times.

OUTSIDE SPRINKLER SYSTEMS, the water curtain is for exposed protection.

AUTOMATIC SPRINKLER SYSTEMS extinguished or held in check 96% of fires in which they were involved.

SPRINKLER PATTERN will equal **16 FOOT DIAMETER** circle at a point 4 feet below the sprinkler head at 15 GPM.

Sprinkler head areas will **OVERLAP**.

Sprinkler systems are to be inspected **4 TIMES PER YEAR**.

Maximum number of sprinkler heads that any be supplied through a single Deluge valve is **5 SPRINKLERS** per 1 1/2 inch valve.

PRE-ACTION SPRINKLER SYSTEMS ARE ACTIVATED BY:
 1. Smoke detectors.
 2. Heat sensors.

SSU = standard upright sprinkler head.

"PINTLE" on sprinkler head indicates that the orifice is smaller than the standard 1/2 inch size.

ACCELERATORS and **EXHAUSTERS** speed up the expelling of air from Dry pipe systems.

OS&Y VALVE:
 1. Outside Screw and Yoke Valve.
 2. Outside Stem and Yoke Valve.

TYPES OF STANDPIPE SYSTEMS:
1. Wet, supply valve open with water pressure at all times.
2. Dry, no permanent water supply in sprinklers.
3. Automatic supply system, opening hose valve.
4. Manual to remote system, is at hose station.

CLASS I STANDPIPE SYSTEM:
For use by Fire Departments and those trained in handling heavy fire streams. (2 1/2 inch hose inside building or for exposure.)

CLASS II STANDPIPE SYSTEMS:
For use primarily by the building occupants until the Fire Department arrives. (1 1/2 inch hose for incipient fires.)

CLASS III STANDPIPE SYSTEMS:
For use by either Fire departments and those trained in handling heavy hose streams or by the building occupants on small hose streams. (for large or small fires.)

GENERAL KNOWLEDGE/FIRE SERVICE INFORMATION QUESTIONS:

1. A firefighter's mental attitude is of vital importance because:
   A. His efficiency on the job is always visible to the public.
   B. He must keep calm in order to think clearly in emergencies.
   C. Firefighting often calls for executive ability.
   D. Intelligence as well as physical stamina is needed.

   ANSWER = B

2. Firefighting ability, in general, is measured in terms of:
   A. Ability to follow orders.
   B. Physical strength.
   C. Reasoning ability.
   D. Intellectual superiority.

   ANSWER = A

3. In a room filled with smoke the best air is located near the:
   A. Ceiling.
   B. Windows.
   C. Floor.
   D. Doors.

   ANSWER = C

4. Which of the following, best describes the purpose of fire doors:
   A. Prevent panic.
   B. Prevent fires.
   C. Prevent arson.
   D. Confine fire to specific section of building.

   ANSWER = D

5. Of the following which is the best conductor of heat:
   A. Iron.
   B. Gas.
   C. Wood.
   D. Air.

   ANSWER = A

6. The main reason that sprinkler systems can, sometimes, secure better fire protection than by a guard is that:
   A. When sprinklers are installed they are in always in position to be activated.
   B. When a guard spots a fire he cannot extinguish the fire as efficiently.
   C. The sprinklers will activate when the room temperature rises slightly.
   D. Sprinklers will activate at periodic intervals.

   ANSWER = A

7. Of the following, which is the principal requirement for safe working conditions:
   A. Housekeeping.
   B. Workmanship.
   C. Weather.
   D. Equipment.

   ANSWER = D

8. Many oil fires will present opportunities for a conflagration, therefore firefighters should consider oil fires as :
   A. A firefighters largest adversary.
   B. The hardest type of fire to combat.
   C. A fire that will spread quickly in all directions.
   D. The same as any fire.

   ANSWER = C

9. Of the following what is the best reason for children to be silent during school fire drills:
   A. Children talking will annoy teachers.
   B. Fire drill will take place with less confusion.
   C. To show how well disciplined the children are.
   D. To impress the children with discipline.

   ANSWER = B

10. Of the following, what is the best reason for fire hydrants being broken:
   A. Hydrants are usually unprotected.
   B. Drivers usually cannot see fire hydrants.
   C. Replacement cost of fire hydrants is not very high.
   D. Hydrants are usually located in the center of the block and they are unprotected.

   ANSWER = A

11. Of the following, what is the best reason for firefighters to clean and check their apparatus and equipment after fires:
   A. So they will have more time for recreation.
   B. So they will look busy.
   C. To keep equipment and apparatus in good shape.
   D. To keep apparatus and equipment looking good.

   ANSWER = C

12. In order to keep business people willing to follow fire safety rules is:
   A. Give them plenty of time to follow the rules.
   B. Order them to follow the rules.
   C. Make sure that they understand the rules.
   D. If they have a new business.

   ANSWER = C

13. Of the following, the primary function of the fire department is:
   A. Promote good citizenship.
   B. Preserve life and property.
   C. Maintain city tranquility.
   D. Reduce business dissension.

   ANSWER = B

14. If a fire is of incendiary origin, it means that:
   A. In its first stage.
   B. A conflagration.
   C. It has been set intentionally.
   D. In its second stage.

   ANSWER = C

15. Of the following, the best reason for using a small nozzle and hose line for an interior apartment fire is:
A. They are more efficient.
B. Large hose is to bulky.
C. They are put in service faster.
D. They will create less water damage.

ANSWER = D

16. Of the following, the best reason for fire departments to have various size ladders is:
A. Fires occur at varying heights.
B. Firefighters are more agile than the general public.
C. Firefighters are taller than the general public.
D. Buildings are not fireproof.

ANSWER = A

17. Of the following the best way for a firefighter to find his way out of a smoke filled building is to:
A. Crawl on the floor with a flashlight.
B. Follow the wall until there is an opening.
C. Follow his hoseline out of the building.
D. Go in the direction of outside sounds.

ANSWER = C

18. Of the following the best way to slow down the spread of fire from one room to another is to:
A. Close the doors between rooms.
B. Remove contents of rooms.
C. Close all the windows in the rooms.
D. Put holes in the walls between rooms.

ANSWER = A

19. When fighting a fire in an un-sprinklered cellar, of the following, which would be the first position of firefighting?
A. Ventilation.
B. A direct fire attack.
C. To prevent the upward spread of fire.
D. Protection of exposures.

ANSWER = C

20. Of the following, which is the greatest danger of a partition fire?
    A. Gases from the fire.
    B. Explosion of the confined gases.
    C. Falling plaster.
    D. The fire rekindling.

    ANSWER = D

21. Fires during high humidity conditions tend to produce, which of the following:
    A. Smoky fires.
    B. Fires in which the smoke will rise very fast.
    C. Fires that are very easy to fight.
    D. Very intense fires.

    ANSWER = A

22. Of the following, which is usually an sign that there are large amounts of CARBON MONOXIDE present at a fire:
    A. Small quantities of smoke.
    B. Large quantities of dense black smoke.
    C. Large quantities of heat and flames.
    D. Small amounts of grayish-blue smoke.

    ANSWER = B

23. Of the following, the most vital factor in the determination the size of hose lines at a fire is:
    A. Size of structure involved in fire.
    B. Height of structure involved in fire.
    C. Square footage of exposures.
    D. Magnitude of the fire and combustibility of the material that is on fire.

    ANSWER = D

24. Of the following, what action should be taken if after making a hole in the roof for ventilation there is only a small amount of draft set up:
    A. Probe the hole with a pike-pole for obstructions.
    B. Make the opening larger.
    C. Check for smoke coming out of other openings.
    D. Make another opening in another area.

    ANSWER = A

25. Of the following, usually, fires in the lower stories of multi-story hotel buildings are:
A. The result of substandard wiring.
B. The result of substandard heating systems.
C. More dangerous to inhabitants of upper floors.
D. More dangerous to inhabitants of lower floors.

ANSWER = C

26. Of the following, which reason is the one that most often motivates an arsonists to set a fire:
A. To hide a crime.
B. To gain profit.
C. To gain revenge.
D. Pyromania.

ANSWER = B

27. What may be assumed from a fire that is spreading too fast for the type of structure?
A. Arson.
B. Quick burning contents.
C. Presence of a substance that will accelerate a fire.
D. All of the above.

ANSWER = D

28. Of the following wood frame roofs, which type presents the greatest fire problem:
A. Pitched.
B. Mansard.
C. Flat.
D. Gambrel.

ANSWER = A

29. Of the following, which is the most important factor to a firefighter concerning a building, from a structural point of view:
A. The strength of the floors.
B. The strength of the supporting beams.
C. Buildings capacity to carry the maximum load for which it was designed.
D. The fact that in a fire the capacities of the normal stress limits will be altered due to heat.

ANSWER = D

30. Of the following, which material is the LEAST resistant to the spread of fire:
    A. Glass blocks.
    B. bricks.
    C. Reinforced concrete.
    D. Hollow tile.

    ANSWER = A

31. Of the following, which material is the most likely to crack when subjected to the rapid heating of fire and the sudden cooling from hose streams:
    A. Bricks.
    B. Heavy timber.
    C. Cast iron.
    D. Steel.

    ANSWER = C

32. Of the following, which is the BEST reason for buildings to have fire-resistive construction:
    A. To lower the chances of a fire starting.
    B. To make a building easier to rebuild after a fire.
    C. To slow the spread of fire.
    D. To protect the building from exposure fires.

    ANSWER = C

33. Of the following, which is the most highly toxic refrigerating agent:
    A. Sulphur dioxide.
    B. Methyl chlorate.
    C. Methyl formate.
    D. Methyl bromide.

    ANSWER = A

34. Of the following, which type of fire is it recommended that firefighters DO NOT use water to extinguish:
    A. Plastic.
    B. Aluminum.
    C. X-Ray film.
    D. All of the above.

    ANSWER = B

35. Of the following choices, which one is the most susceptible to spontaneous ignition:
    A. Rags saturated with motor oil.
    B. Rags saturated with castor oil.
    C. Rags saturated with fuel oil.
    D. Rags saturated with linseed oil.

    ANSWER = D

36. Of the following, which choice is the best reason for smoldering fires to have high percentages of carbon monoxide:
    A. The high temperatures.
    B. The low temperatures.
    C. The high amounts of oxygen.
    D. The low amounts of oxygen.

    ANSWER = D

37. When the oxygen concentration of the air is lower than 16 per cent:
    A. Breathing is not affected.
    B. Most substances will continue to burn freely.
    C. Most flames are extinguished.
    D. No fire will continue burning.

    ANSWER = C

38. Of the following materials, which one will give off oxygen most readily when heated:
    A. Potassium nitrate
    B. Potassium chlorate.
    C. Calcium oxide.
    D. Sodium nitrate.

    ANSWER = B

39. Of the following substances, which one will usually give off a white smoke when burning:
    A. Gun powder.
    B. Gasoline.
    C. Vegetable compounds.
    D. Nitrocellose film.

    ANSWER = C

40. When the temperature of a fire increases, the amount of carbon monoxide in the gases released:
   A. Increases.
   B. Decreases.
   C. Stays constant.
   D. Varies.

   ANSWER = A

41. By increasing the area of contact, wetting agents added to water:
   A. Increase the surface tension.
   B. Increase the rate of heat transfer.
   C. Decrease the rate of heat transfer.
   D. Decrease the penetrating absorption.

   ANSWER = B

42. Of the following, the BEST reason that small fire streams are often of little value in fighting very hot fires is that:
   A. Small streams may vaporize prior to reaching the fire.
   B. Small streams will break-up into hydrogen and oxygen thus increasing the heat.
   C. The small streams will not allow the firefighter to get close to the seat of the fire.
   D. None of the above.

   ANSWER = C

43. Of the following, which type of fire are wetting agents and plain water the MOST effective in:
   A. Class A fires.
   B. Class B fires.
   C. Class C fires.
   D. None of the above.

   ANSWER = A

44. Of the following, the BEST reason that the exterior standpipes should be drained after use is to:
   A. Relive them of the load of the water.
   B. Make their future operation easier.
   C. Prevent the buildup of rust.
   D. Prevent the collection of algae.

   ANSWER = A

45. Of the following incipient fires, which one would water fire extinguisher be the LEAST effective on:
   A. Textiles.
   B. Greases in open containers.
   C. Paper.
   D. Wood.

   ANSWER = B

46. Of the following fires, which one are foam fire extinguisher the LEAST effective on:
   A. Flammable liquids contained in a small area.
   B. Wood and/or paper.
   C. Flammable liquid fire on a wooden floor.
   D. Grease fire.

   ANSWER = B

47. Of the following fire extinguisher, which one would be the MOST useful on an electrical fire:
   A. Foam.
   B. Water.
   C. Dry chemical.
   D. Vaporizing liquid.

   ANSWER = C

48. Of the following, which is the greatest cause for the loss of firefighters lives:
   A. Lack of proper drills.
   B. Lack of equipment.
   C. Lack of manpower.
   D. Lack of knowledge of the local structural conditions.

   ANSWER = D

49. Of the following, which one is the LARGEST single cause of deaths due to fire and rapid spread of fire in older buildings:
   A. Faulty egress.
   B. Faulty building construction.
   C. Improper fire escapes.
   D. Unprotected vertical openings.

   ANSWER = D

50. Of the following, which is the BEST method to reduce the fire hazard of a Christmas tree:
A. Notch the trunk of the tree.
B. Char the cut end of the tree.
C. Keep the cut end of the tree in water.
D. Put wax on the cut end of the tree.

ANSWER = C

SECTION 12

AFTER THE EXAM

FOR FUTURE REFERENCE

Upon completion of the **WRITTEN EXAM** make sure that you hand in all of the test matter to the proctor.

After the exam you may ask the proctor as to when you will be notified of the exam results.

After you leave the test location, you should go somewhere that you are comfortable and take the time to list of:

1. The type of questions that were on the exam, ie: multiple choice, true or false, matching, fill-in, etc.

2. The areas that were tested, ie: general knowledge, figures, grammar, mechanical, mathematical, etc.

3. Any question/questions that you feel that you should save on a continual ever growing list of test questions for your future reference.

4. Anything or feeling that you have about this particular exam that you might want to recall at a later date.

5. Any ideas that you may have on how you could have improved your preparation for this exam.

Editors note: this text includes excerpts from the following Books written by the author and available from Information Guides Dept. "B", P.O. Box 531, Hermosa Beach, CA 90254:

A SYSTEM FOR ADVANCEMENT IN THE FIRE SERVICE

FIREFIGHTER WRITTEN EXAM STUDY GUIDE

FIREFIGHTER ORAL EXAM STUDY GUIDE

FIRE ENGINEER WRITTEN EXAM STUDY GUIDE

FIRE ENGINEER ORAL EXAM STUDY GUIDE

FIRE CAPTAIN WRITTEN EXAM STUDY GUIDE

FIRE CAPTAIN ORAL EXAM STUDY GUIDE

INDEX

INDEX

"A"

ACROLEIN	278
ADDITION	
PROBLEMS	183
ADIABATIC	171
AERIAL LADDERS	260
BOOMS	261
AIR	276, 277
AIR CLEANER	114
ALPHABETICAL	
PROGRESSIONS	219
ALTERNATOR	112
FLUCTUATING	111
AMMETER	111
AMMONIA	278
AMPERES	5
ANGLE	169
APPARATUS	114
APTITUDE	2
AREAS	
VOLUMES	120
ARITHMETIC	182
ARSON	274
ARTIFICIAL VENTILATION	291
ATMOSPHERIC	
PRESSURE	165
AUTOMATIC SPRINKLER SYSTEMS	300

"B"

B.T.U.	59
BRITISH THERMAL UNIT	159
BACKDRAFT	276, 289
BAROMETRIC PRESSURE	253
BATTERIES	110, 111
CHARGE	111
CHARGING RATE	110
ELECTROLYTE	110
OVERCHARGING	111
SPECIFIC GRAVITY	110
SULFURIC ACID	110
BATTERY	111
OVERCHARGING	111
BELTS	119
BIMETALLIC	169
BLEVE:	282
BLOCKS	
MATCHING	224
BOILING POINT	160
BOOKS	
A SYSTEM FOR ADVANCEMENT IN THE FIRE SERVICE	315
FIREFIGHTER ORAL EXAM	315
FIREFIGHTER WRITTEN EXAM	315
FIRE CAPTAIN ORAL EXAM	315
FIRE CAPTAIN WRITTEN EXAM	315
FIRE ENGINEER WRITTEN EXAM	315
BOYLE'S LAW	292

BRAKES	110
EXCESSIVE HEAT	110
FADING	110
FLUID	110
LINING	110
BREATHING APPARATUS	
SCBA	109
BRITISH THERMAL UNIT	59
B.T.U.	159
BUILDING CONSTRUCTION	270
BURNS, CHEMICAL	205

"C"

CAM FOLLOWER	113
CAM SHAFT	113
CAMBER	116, 117
CAMBER	113
CANTILEVER	251
CAPACITOR	112
CARBON BLACK	114
CARBON DIOXIDE	277, 278
CARBON MONOXIDE	61, 276
CARBURETOR	112, 114
CARRYALL	269
CASTER	113
CAVITATION	258
CENTRIFUGAL	254
FORCE	168
PUMPS	255
CHEMICAL	
REACTIONS	162, 281
CHEMICALS:	
COMBUSTIBLE	281
CORROSIVE	282
UNSTABLE	281
CHEMTRIC	283
CHLORINE	278
CHOKE	113, 114
CIRCUMFERENCE	169
CLUTCH	113
COIL	112
COMBUSTIBLE	
CHEMICALS	281
LIQUIDS	161
METALS	282
COMBUSTION PRODUCTS	289
COMMAND POST	286
COMPOUND GAGES	253
COMPRESSION	113, 115
COMPRESSION GAUGE	112
CONDENSER	112
CONDUCTOR	170
CONFLAGRATION	290
CONTUSION	204
RADIATOR, COOLER	254
CORROSIVE CHEMICALS	282
CUBES, MATCHING	224
CURRENT REGULATOR	111

"D"

D.O.T.	283, 284
DIESEL	115
ENGINES	115
TURBOCHARGED	115
DISPLACEMENT	253
DISTRIBUTOR	112
DIVISON PROBLEMS	189
DOORS, EXIT	272
DOORS	271
DRIVING	247, 248
BLIND SPOT	248
BRAKING	247
CODE 3	247
DEFENSIVE	248
PERCEPTION	247
REACTION	247
ROAD SIGNS	247
STOPPING	247
TRAFFIC SIGNS	247
DUST	283
DWELL ANGLE	112

"E"

E.D.I.T.H.	274
EGRESS	272
ELECTRICAL	5
ELECTRICAL SHOCK	107
ELECTRICITY	105
ELECTRON	105
ELEMENTS	171
EMT-I	202
EMT-I	
INFORMATION	208
QUESTIONS	211
TOPICS	209
ENDOTHERMIC REACTION	283
ENGINES	
4 STROKE	116
DIESEL	115
GASOLINE	115
EQUIPMENT	108
EXAM, MECHANICAL	104
EXAM SUGGESTIONS	6
EXAMS	
CHEMISTRY	158
ESSAY QUESTIONS	5
MATCHING	5
MULTIPLE CHOICE	3
PHYSICS	158
SCIENCE	158
TRUE-FALSE	4
TYPES OF EXAMS	3
VOCABULARY	78
EXCEEDED	257
EXIT	272
EXPOSURES	291
EXTENSION LADDERS	250

EXTINGUISHER	45, 296
CARBON DIOXIDE	297, 298
DRY CHEMICAL	296, 297
DRY POWDER	297
FOAM	43, 297
HALON	47
MULTIPURPOSE	296, 297
OBSOLETE	298
SODA ACID	296
WATER	293
LIQUIFIED	47
EXTINGUISHING SYSTEMS	293

"F"

FIGURES, MATCHING FORMS	230
FIRE	
CAUSE	273
CLASSIFICATIONS	28
DANGER	290
DEVIL	290
EXTENSION	291
EXTINGUISHER	293
FLOOR	286
FLOW	287
HAZARD	273, 274
HOSE	30, 32, 265
HYDRAULICS	165
LOAD	270
LOSS	58
MODES	289
POINT	161
PUMPS	253
SCENE	288
SCIENCE	288
SPREAD	288
STORM	288
STRATEGY	49, 288
STREAMS	261
TACTICS	288
TETRAHEDRON	288
TRIANGLE	60, 288
VENTILATION	291
WIND	289
FIRE BEHAVIOR	285
FIRE EXT. RATINGS	293
FIRE PREVENTION	270
FIRE SERVICE INFORMATION	245
DRIVING	247
LADDERS	250
FIRE STREAMS	
SPRAY	295
FIRE	
FLOOR	287
MODES	24
STAGES	289, 24
FIRE-RESISTIVE	271

"F" (continued)

```
FIREBREAK ....................... 290
FIREFIGHTER DUTIES ............... 27
FIREFIGHTING .................... 285
         BLITZ ATTACK .......... 287
         INDIRECT ATTACK ....... 286
         SIZE-UP ............... 288
         SIZE-UP: .............. 287
FIRES
         METAL ................. 298
BASEMENT ........................ 292
FIRST AID ....................... 202
         INFORMATION ........... 202
         QUESTIONS ............. 203
FITTINGS ........................ 264
FLAMMABLE
         DENSITY ............... 161
         LIQUIDS ............... 161
FLASH POINT ................. 56, 160
FLASHOVER .................... 5, 288
FOAM ....................... 294, 295
         AFFF .................. 295
         ALCHOL ................ 296
         COMVENTIONAL .......... 295
         HIGH EXPANSION ........ 295
         LOW EXPANSION ......... 296
         ORDINARY .............. 296
         PROPORTIONER .......... 296
FOG STREAM ...................... 263
FORCIBLE ENTRY .................. 268
FORMS, MATCHING ................. 224
FRACTURE, COMPOUND .............. 204
FREON ........................... 278
FRICTION LOSS ............... 165, 168
FUEL ....................... 114, 115
         MIXTURE ............... 115
         OCTANE ................ 114
FULCRUM ......................... 168
FUSES ........................... 112
```

"G"

```
GAGES, COMPOUND ................. 253
GASES, L-P ...................... 279
GASOLINE ENGINES ................ 115
GAWR. ..................254 ...... 254
GEAR ............................ 113
GENERAL KNOWLEDGE ............... 245
GENERATOR ....................... 111
GOVERNOR ........................ 253
GRAMMAR .......................... 93
         EXAM .................. 93
         TEST .................. 93
         PUNCTUATION ........... 94
         USAGE ................. 94
GRAVITY, SPECIFIC ............... 160
GVWR ............................ 254
```

"H"

```
HALOGENATE ...................... 298
HALOGENS ........................ 279
HALON ........................... 299
HAZARD
         CONFLAGRATION ......... 290
         FIRE .............. 273, 274
         LIGHT ................. 45
         TARGET ................ 274
HAZARDOUS MATERIALS ............. 276
HAZARDS
         OCCUPANCY ............. 273
         EXTRA ................. 45
         ORDINARY .............. 45
HEAD PRESSURE .............. 166, 256
HEAT
         EXCHANGER ............. 264
         EXHAUSTION ............ 204
         LATENT ................ 159
         SPECIFIC .............. 159
         STROKE ................ 211
ENDOTHERMOIC .................... 290
         EXOTHERMIC ............ 290
         TRANSMISSION .......... 290
HORIZONTAL REACH ................ 262
HOSE ............................ 265
         CLAMPS ................ 108
         STREAMS ............... 261
HURST TOOL ...................... 38
HYDRAULIC FLUID ................. 251
HYDRAULICS ...................... 165
HYDROGEN ........................ 278
         CYANIDE ............... 277
         SULFIDE ............... 277
```

"I"

```
IGNITION ................... 112, 115
         SPONTANEOUS ....... 41, 280
         TEMPERATURE ........... 161
         COIL .................. 112
         SWITCH ................ 112
INFORMATION
         EMT-I ................. 208
         FIRE CAPTAIN .......... 317
         FIRE ENGINEER ......... 315
         FIREFIGHTING .......... 285
         FIRST AID ............. 202
         GENERAL ............... 105
         MECHANICAL ............ 110
         PHYSICS ............... 168
         FIRE SERVICE .......... 159
INITIAL RESPONSE ................ 287
INSPECTION ...................... 273
INSPECTION, SCHOOL .............. 274
INSULATOR ...................170
```

JAWS OF LIFE 38

"K"

KNOTS 152

"L"

LABELS, D.O.T. 283
LACERATION 207
LADDER 251
LADDER RAISING 55, 251
LADDER COMPANY 250
LADDER TRUCK 251
LADDER TRUCKS 250
LADDERS 250, 252
    AERIAL 250
    CAPACITIES 250
    EXTENSION 250
    N.F.P.A 193 250
    SIZE 250
    TRUSS 250
LEEWARD 269
LEVERAGE 145, 168
LIFE SAFETY CODE 273
LIGHT 170
LIQUIDS: COMBUSTIBLE 161, 275
    FLAMMABLE 161, 274
LUBRICANTS 116
LUGGING 249

"M"

MAGNETIZATION 169
MAGNETO 112
MANPOWER 288
MASTER STREAM 265
MATCHING: KNOTS 152
    MECHANICAL PARTS 151
    TOOLS 150
MATCHING FORMS 224
    BLOCKS 224
    CUBES 224
    FIGURES 230
MATH: ADDITION 183
    CONCEPTS 182
    DIVISION 189
    MULTIPLICATION 187
    PERCENTAGE 191
    SUBTRACTION 185
    WORD PROBLEMS 193
MECHANIC: COMPREHENSION 104
    INFORMATION 110
    OBJECT QUESTIONS 153
    PRINCIPLES 119
    QUESTIONS 104, 145
    TEST 104
METALS, COMBUSTIBLE 282
MULTIPLICATION PROBLEMS 187
MUSHROOMING 292

"N"

N.F.P.A. 270, 271
N.F.P.A. #196 266
N.F.P.A. #198 266
N.F.P.A. 1901 266
N.F.P.A. STANDARD #10 293
N.F.P.A., ASSOCIATION 270
NET PUMP PRESSURE 254
NEUTRONS 106
NITRATES .... 281 ... NITRITES 281
NITROGEN DIOXIDE 278
NOZZLE: PRESSURE 166
     REACTION 166
NUMERICAL PROGRESSION 220

"O"

OCCUPANCIES, CLASSIFICATION 273
OCTANE 115
OHMS 5
OIL 112, 114, 116
    DILUTION 112
    GAUGE 112
    S.A.E 112
    VISCOSITY 112
OPENINGS 272
ORDERS, INCONSISTENT 10
OS&Y VALVE 300
OVERHAUL 268
OXIDATION 279, 280

"P"

PACKING 267
PANIC HARDWARE 272
PARALLEL 256
PERCENTAGE PROBLEMS 191
PH SCALE 284
PHYSICS 168
PISTON 112
PITOMETER: 109
POISONING, CARBON MONOXIDE ... 203
POST INDICATOR VALVE
    PIV 299
PRE-FIRE PLANS 287
PRE-TEST MANUALS 3
PRESSURE 257, 283
    BACK 166
    KINETIC 171
    VAPOR 161
PRESSURE 166
    ATMOSPHERIC 165
    NOZZLE 166
PROGRESSIONS
ALPHABETICAL 219
    NUMERICAL 218, 220
FORMS 218
PROPORTIONER 296
PROTONS 106
PULLEYS 119

"P" (continued)

PUMP
- CENTRIFUGAL 264
- CERTIFICATION TEST 268
- DELIVERY TEST 269
- IMPELLER: 269
- MULTI-STAGE 269
- POSITIVE DISPLACEMENT 263, 265
- SERVICE TEST 269

PUMPER 54, 113, 264
- QUAD 264
- QUINT 264
- TRIPLE 264
- TRIPLE COMBINATION 54

PUMPER SUBTRACTION 54

PUMPS 267
- ACTION 265
- CAPACITIES 54
- CENTRIFUGAL 51, 254-256
- DISPLACEMENT 253
- POSITIVE 263-265
- POSITIVE-CAPACITY 253
- PRINCIPLE 264
- RECIPROCATING 253
- TESTING 254
- BOOSTER 258

"Q"

QUAD 264
QUESTIONS
- ADDITION 183
- DIVISION 189
- EMT-I 211
- FIRST AID 203
- MATCHING FORMS 225
- MECHANICAL 145, 153, 187
- MULTIPLICATION 187
- PERCENTAGE 191
- READING 22
- SCIENCE/CHEMISTRY/PHYSICS 172
- SITUATION 8
- SPELLING 82
- SUBTRACTION 185
- PULLEYS/GEARS/LEVERS 121

QUINT: 264

"R"

RADIATOR 112
RADIATOR COOLER 264
RADIUS 169
READING COMPREHENSION 22
RECTIFIER 112
REFERENCE, FUTURE 314
RELIEF VALVE 253
RELIEF VALVES 253
ROAD SIGNS 247

"S"

SALVAGE COVERS 268, 269
SCBA 109
SCIENCE EXAMS 158
SEIZURE 207
SHOCK 205
SIAMESE 264
SIZE-UP 42, 288
SMOKE DETECTORS 274
SMOKES DEADLY TRIO 276
SPECIFIC GRAVITY 22, 164
SPEEDOMETER 113
SPELLING QUESTIONS 82
SPRAY STREAM 263
SPRINKLER SYSTEMS
- DELUGE 300
- DRY PIPE 300
- PRE-ACTION 300
- RATINGS 299
- WATER SUPPLY 299
- WET PIPE 300
- OUTSIDE 300
- AUTOMATIC............ 299

STANDPIPE SYSTEMS
- TYPES 301
- CLASS I 301
- CLASS II 301
- CLASS III 301

STEAM. 294
STRAIGHT LADDERS 250
STRAIGHT STREAM 263
STRATEGY, FIRE 288
STREAM, HEAVY FIRE 286
SUBJECT MATTER 2
SUBLIMATION 280
SUBLIME 159
SUBTRACTION PROBLEMS 185
SURFACE TENSION 294

"T"

TACHOMETER 113
TACTICA, FIRE 288
TAMPONADE, PERICARDIAL 211
TARGET HAZARD 271, 274
TASK FORCE 288
TEAMWORK 55
TEST: APTITUDE 8
- MECHANICAL 104
- VOCABULARY 76
- SPELLING 92
- VOCABULARY 65

TETRAHEDRON. 5
THERMOSTAT 112, 113, 169
TILLERMAN 117
TIRE WEAR: UNBALANCE 116
- TOE-IN...........117....

TOOLS 108, 268
- TOOLS: MATCHING QUESTIONS 150

TORQUE ....113 ....
- TRANSFER VALVE 257
- TRANSMISSION 113
- TRAUMA 211
- TRIPLE COMBINATION PUMPER 264

322

"U"

UNIFORM BUILDING CODE
    U.B.C. 270
UNIFORM FIRE CODE
    U.F.C. 270
UNSTABLE CHEMICALS 281

"V"

VACUUM 165
VALVE: TRANSFER 257
VALVE LIFTER 112
VAPOR 159
VAPOR PRESSURE 161
VAPOR DENSITY 23, 160, 164
VAPOR LOCK 115
VELOCITY 256, 257
VENTILATION 11, 37, 268, 291, 292
VERBAL ABILITY 65
VERTICAL REACH 262
VOCABULARY 65
VOCABULARY TEST 65, 76
VOLTAGE 5
VOLTAGE REGULATOR 112
VOLUTE 255, 260

"W"

WALLS 272
WATER ... 62, 167, 285, 286, 293, 294, 296
    COOLING EFFECT 62
    HAMMER 34
    LIGHT 295
    VISCOUS 295
WEATHER 290
WET-WATER 294
WETTING AGENTS 294
WHEELS 119
WINDWARD 269
WYE 263

FIRE SERVICE STUDY GUIDES

FOR

FIRE CAPTAIN

FIRE LIEUTENANT

FIRE ENGINEER

FIRE PUMP OPERATOR

FIRE APPARATUS DRIVER

ENTRANCE LEVEL FIREFIGHTER

ADVANCEMENT IN THE FIRE SERVICE

FIRE SERVICE EXAM CLASSIFICATIONS:

BACKGROUND INVESTIGATIONS

PHYSICAL AGILITY EVENTS

MEDICAL EXAMINATIONS

ASSESSMENT CENTERS

ORAL INTERVIEWS

SIMULATIONS

PRACTICAL

WRITTEN

ABSOLUTELY GUARANTEED!

The information contained in these **STUDY GUIDES** will give **YOU** the **POSITIVE EDGE** in the **FIRE SERVICE EXAM** process. These **STUDY GUIDES** are **ORGANIZED** so that **YOU** can cover a maximum amount of **VARIED** information without having to spend **VALUABLE** study time searching for the required information.

YOU WILL COVER MORE INFORMATION IN LESS TIME! IF NOT FULLY SATISFIED RETURN BOOK/BOOKS FOR FULL REFUND !

BOOKS AVAILABLE FROM INFORMATION GUIDES ARE LISTED ON THE FOLLOWING PAGES, ALONG WITH AN ORDER FORM:

BOOK #1
FIRE ENGINEER WRITTEN EXAM

Fire apparatus, fire prevention, fire pumps, fire streams, fire behavior, fire hydraulics, fire prevention, tools and equipment, water supply, hazardous materials, fire extinguishing systems.

BOOK #2
FIRE ENGINEER ORAL EXAM

Oral interview preparation, job knowledge, personal information general knowledge, and actual situation questions and responses.

BOOK #3
FIRE CAPTAIN WRITTEN EXAM

Fire administration, training, fire fighting fire prevention, fire behavior, fire apparatus and equipment, fire extinguishing systems, fire streams, water supply, and hazardous materials.

BOOK #4
FIRE CAPTAIN ORAL EXAM

Oral interview preparation, job knowledge, personal information, general knowledge, situation questions, incident simulator preparation, assessment centers.

BOOK #5
COMPLETE FIREFIGHTER CANDIDATE

Introduction to the Fire Service, Fire Department familiarization, exam check list, practice exams, locating exams, job announcements, job applications, resumes, exam process, exam divisions, what to prepare for, written/oral/agility/medical exam information and much more.

BOOK #6
FIREFIGHTER WRITTEN EXAM

Types of exams and questions, general aptitude and judgement, reading, vocabulary, spelling, grammar, science, mechanical comprehension, physics, chemistry, pattern analysis, math, progressions, Fire Service information, first aid, pre-test study books, and much more!

BOOK #7
FIREFIGHTER ORAL EXAM

Oral interview preparation, job knowledge, general knowledge, personal knowledge, actual situation questions and responses.

BOOK #8

ADVANCEMENT IN THE FIRE SERVICE
Prior to obtaining a position as a Firefighter, after obtaining a position as a Firefighter, after promoting to the position of Engineer, after promoting to the position of Captain.

BOOK #1

"FIRE ENGINEER WRITTEN EXAM STUDY GUIDE" 2nd. ed.
$15.95 per copy

180 page book with 10 chapters containing over **2500** selections of information that **ALL FIRE FIGHTERS** should know. Each piece of information is presented in a **NEW** and **UNIQUE METHOD** that will permit **EFFICIENT - ORGANIZED** studying. A must book that will assist **ANY FIREFIGHTER** in obtaining a high score on the **WRITTEN** portion of the **FIRE ENGINEERS** promotional exam.
Perfect bound soft cover: 5 1/2" X 8 1/2"
ISBN 0-938329-52-9   LCCN 86-81239

BOOK #2

"FIRE ENGINEER ORAL EXAM STUDY GUIDE" 2nd ed.
$15.95 per copy

192 page book with seven chapters containing over **400 PRACTICAL - ORAL INTERVIEW** and **SITUATION** type questions. Each question is followed by a **SUGGESTED** response. The **MOST COMPLETE** book of it's kind. An **ABSOLUTE MUST** as a tool to prepare **ANY FIRE FIGHTER** for the **PRACTICAL - ORAL** portion of the **FIRE ENGINEER - APPARATUS DRIVER - PUMP OPERATOR** promotional exam.
Perfect bound soft cover: 5 1/2" X 8 1/2"
ISBN 0-938329-53-7   LCCN 88-80888

BOOK #3

"FIRE CAPTAIN WRITTEN EXAM STUDY GUIDE"
$18.95 per copy

288 page book with ten chapters containing over **3000** selections of information **ALL FIRE FIGHTERS** should know. Each bit of information is presented in a **NEW/UNIQUE METHOD** that will permit **EFFICIENT** and **ORGANIZED** studying. This is a **MUST BOOK** that will assist **ANY FIRE FIGHTER** in obtaining a high score on the **WRITTEN** portion of the **FIRE CAPTAIN - LIEUTENANT** promotional exam.
Perfect bound soft cover: 5 1/2" X 8 1/2"
ISBN 0-938329-54-5   LCCN 88-80890

BOOK #4

"FIRE CAPTAIN ORAL EXAM STUDY GUIDE"
$18.95 per copy

228 page book with ten chapters containing over **500 ORAL INTERVIEW** and **SITUATION** type questions. Each question is followed by a **SUGGESTED** response. The most complete book of its kind. An **ABSOLUTE MUST**, as a tool that will prepare **ANY FIRE FIGHTER** for the **ORAL** portion of the **FIRE CAPTAIN - LIEUTENANT** promotional exam.
Perfect bound soft cover: 5 1/2" X 8 1/2"
ISBN 0-938329-55-3   LCCN 88-80889

BOOK #5

"THE COMPLETE FIREFIGHTER CANDIDATE"
$12.95 per copy

150 page book with eight chapters containing the **MOST COMPLETE** inventory of information available. This book will guide prospective Firefighters through all the essential steps that need to be taken in order to become the **COMPLETE FIREFIGHTER CANDIDATE**.

Perfect bound soft cover   5 1/2" X 8 1/2"
ISBN 0-938329-58-8   LCCN 89-81738

BOOK #6

"FIREFIGHTER WRITTEN EXAM STUDY GUIDE"
$19.95 per copy

336 page book with **12** chapters containing over **3000** selections of information - questions - answers that **ALL PROSPECTIVE FIREFIGHTERS** should know. Each selection of information is presented in a **UNIQUE METHOD** that will permit **EFFICIENT** and **ORGANIZED** studying. A must book that will assist Firefighter candidates in obtaining a high score on the **WRITTEN** portion of the **FIREFIGHTERS ENTRANCE EXAM**.

Perfect bound soft cover: 5 1/2" X 8 1/2"
ISBN 0-938329-59-6   LCCN 89-81736

BOOK #7

"FIREFIGHTER ORAL EXAM STUDY GUIDE"
$15.95 per copy

226 page book with eight chapters containing over **400 ORAL INTERVIEW** and **SITUATION** type questions. Each question is followed by a suggested response. The **MOST COMPLETE** book available for preparing Firefighter candidates for the **ORAL** portion of the **FIREFIGHTERS ENTRANCE EXAM**.

Perfect bound soft cover: 5 1/2" X 8 1/2"
ISBN 0-938328-61-8   LCCN 89-81737

BOOK #8

A SYSTEM FOR: " ADVANCEMENT IN THE FIRE SERVICE"
$9.95 per copy

170 page book with six chapters containing all the essential steps and procedures that Firefighters should follow in order to promote within the **FIRE SERVICE**. This book is presented in a **THOROUGH** and **ORGANIZED** manner so as to allow Firefighters to see the overall picture for **ADVANCEMENT IN THE FIRE SERVICE**.

Perfect bound soft cover: 5 1/2" X 8 1/2"
ISBN 0-938329-56-1   LCCN 88-083385

BOOK REVIEWS OF "STUDY GUIDES"

FIREFIGHTER NEWS:

"Every Firefighter that has been or is currently involved in the Fire Service exam process will attest to the need for these books."

REKINDLE MAGAZINE:

"Information is presented in a new and unique method for efficient and organized studying."

FIREHOUSE MAGAZINE:

"Well thought-out study books for the person aspiring to promote."

FIRE CHIEF MAGAZINE:

"Information that Firefighters - Candidates should know in preparing for Fire Service exams."

FIRE TRAINING OFFICER:

"I used the books as a reference for a portion of our Fire Department exams, very useful."

GRANVILLE, TEXAS - FIREFIGHTER:

"There is a need for more books of this type!"

FIRE ENGINEERS JOURNAL MAGAZINE:

"Stimulates the reader to think about relevant facts presented in a two or three line summary, which can be retained in memory!"

THE INSTITUTION OF FIRE ENGINEERS:

"These books are a benefit to all Firefighters - Candidates taking Fire Service examinations."

FIREFIGHTER ESCONDIDO, CALIFORNIA:

"Excellent books, excellent format, excellent content, very useful."

YUBA CITY - FIRE EQUIPMENT OPERATOR:

"Very useful and informative. Need more books like these."

COLLEGE FIRE SCIENCE INSTRUCTOR:

"Books are very useful as reference guides for all of my classes!"

ORDER FORM

NAME _____

FIRE DEPARTMENT _____

STREET ADDRESS _____

CITY _____

STATE AND ZIP _____

PHONE NUMBER (_____) _____

NUMBER OF BOOKS:

#1 ___ $15.95 = $ _____ . ___   #5 ___ $12.95 = $ _____ . ___

#2 ___ $15.95 = $ _____ . ___   #6 ___ $19.95 = $ _____ . ___

#3 ___ $18.95 = $ _____ . ___   #7 ___ $15.95 = $ _____ . ___

#4 ___ $18.95 = $ _____ . ___   #8 ___ $ 9.95 = $ _____ . ___

CALIFORNIA RESIDENCE ADD 6.5% SALES TAX:
(.065 times the total price of books ordered)

SALES TAX = $ _____

SHIPPING & HANDLING CHARGES:

BOOK RATE MAIL:
$1.00 for the 1st copy; $0.50 each additional copy.

RUSH FIRST CLASS MAIL:
$3.00 for the 1st copy; $1.00 each additional copy.

SHIPPING AND HANDLING = $ _____

MAKE CHECKS or M.O. OUT TO : "INFORMATION GUIDES"

TOTAL AMOUNT ENCLOSED = $ _____

CREDIT CARD ORDERS : _____ VISA _____ MASTER CARD

CREDIT CARD NUMBER : __ __ __ __ __ __ __ __ __ __ __ __ __ __

EXPIRATION DATE : _____

NAME ON CARD : _____

SIGNATURE : _____

RUSH ORDERS PHONE: 1-800 "FIRE BKS" = 1-800-347-3257
ALSO: (213) 379-1094